佳图顶级系列

Top Show Flat III

顶级样板房 III

佳图文化 编

下 册

天津大学出版社
TIANJIN UNIVERSITY PRESS

图书在版编目（CIP）数据

顶级样板房 .3：全 2 册 / 佳图文化编 . — 天津：
天津大学出版社，2013.10
ISBN 978-7-5618-4762-6

Ⅰ.①顶··· Ⅱ.①佳··· Ⅲ.①室内装饰设计—
作品集—中国—现代 Ⅳ.① TU238

中国版本图书馆 CIP 数据核字（2013）第 198144 号

策　　划	王　志
责任编辑	油俊伟
出版发行	天津大学出版社
出版人	杨　欢
地　　址	天津市卫津路 92 号天津大学内（邮编：300072）
电　　话	发行部：022—27403647　邮购部：022—27402742
网　　址	publish.tju.edu.cn
印　　刷	利丰雅高印刷（深圳）有限公司
经　　销	全国各地新华书店
开　　本	245mm×325mm
印　　张	33
字　　数	867 千
版　　次	2013 年 10 月第 1 版
印　　次	2013 年 10 月第 1 次
定　　价	596.00 元（上下册）

Preface 前言

Developing from nothing to refinement, from youth to maturity, show flat is not only a product of rapid development of the housing market, but also a concrete manifestation of the residential culture. Show flat exists not only for showing and sales promotion, but also bears the function of leading the masses living trend. As an important factor in the process of building sales, show flat is getting more and more attention from the real estate developers and popular with many buyers. Therefore, the design of the show flat becomes important and decisive.

Professional show flat design needs the designers to have deep understanding on the popular trend and future residential development, and also know well about the location of the building, cultural customs and the quality of the consumer group, thus they can plan reasonable space functional zones and decide space decoration style from a professional perspective. Generally speaking, top show flat design must be the integration of functionality, decorativeness and culture connotation.

As a continued professional book of the top show flat series, this book selects domestic high-level top showflat cases with rich content, unique design layout and innovative, avant-garde demonstration, so we believe that this book will bring visual enjoyment of beauty and design inspiration for the interior designers and other related professional readers.

样板房从无到有，从有到精，从青涩到成熟，既是楼市快速发展的一个产物，也是住宅文化的一种具体表现。样板房不仅以展示促销为目的，还担负着引领大众居住潮流的功能。作为楼盘销售过程中的一个重要因素，样板房也越来越受到房地产开发商的重视和广大购房者的喜爱。因而,样板房的设计就变得举足轻重了。

专业的样板房设计需要设计师不仅对流行趋势和未来住宅的发展动态有所了解，而且对楼盘的地理位置、人文习俗以及消费群体的品质都有所掌握，然后再从专业的角度去合理规划空间功能区和定义空间装修风格。总的来说，顶级的样板房设计一定是集功能性、装饰性以及文化内涵为一体的。

作为顶级样板房系列中承前启后的专业读本，本书精选国内高水平的顶级样板房案例，通过丰富的内容、独特的设计排版以及新颖前卫的展示，我们有理由相信：本书定会给室内设计师及相关专业的读者以美的视觉享受和设计启迪。

CONTENTS 目 录

Chinese Style
中式风格

- 008　Hung Hom Lingshijun Pulin Garden—Show Flat of Apartment C
  红磡领世郡普霖花园C户型样板房
- 020　MOON RIVER Show Flat 1
  水月周庄样板房一
- 028　Show Flat of Oriental Charm, Changzhou
  常州东方风韵样板房
- 042　Show Flat of Shenzhen Jixin City
  深圳集信名城样板房
- 054　Show Flat of Beijing Yanxitai No. 37
  北京燕西台37号样板房
- 060　Type A1 Show Flat of Junyue Golden Beach, Zhangzhou
  漳州君悦黄金海岸A1户型样板房
- 066　Top Inkstone Residence, Taoyuan
  桃园一品国砚
- 078　Show Flat of Lonyear Villa Experience Hall
  苏州朗誉别墅装饰实景体验馆样板间

Mediterranean Style
地中海风格

- 088　Type A Show Flat of Red Coral Bay, Shijiazhuang
  石家庄红珊湾A户型样板房

Neo-classical Style
新古典主义风格

- 098　Show Flat of Fuzhou Sansheng Central Park
  福州三盛中央公园样板房
- 106　Show Flat of East Shore International Garden, Hohhot
  呼和浩特东岸国际花园样板房
- 110　Type B Show Flat of Coral Bay, Shijiazhuang
  石家庄红珊湾B户型样板房

American Style
美式风格

120　Show Flat of Lake and Villa, Harbin
　　哈尔滨湖与墅样板房

132　Huangshang Paradise City House Type C-2 Show Flat
　　黄山和泰国际城C-2户型样板房

European Style
欧式风格

142　Show Flat of Fuqing Jinhui Huafu
　　福清金辉华府样板房

150　Show Flat of Foshan Vanke Crystal City
　　佛山万科水晶城样板房

158　Show Flat A of Greattown International, Fuzhou
　　福州名城国际A样板房

166　Show Flat B of Greattown International, Fuzhou
　　福州名城国际B样板房

176　Type D Show Flat of Qinyang Rose City
　　沁阳玫瑰城D户型样板房

184　Show Flat of Shenzhen CTS International Mansion
　　深圳中旅国际公馆样板房

194　Show Flat of Vanke Wonder Town, Chongqing
　　重庆万科缇香郡样板房

206　Type B Show Flat of Chongqing Zhonghai International Community No.3 Building
　　重庆中海国际社区3号楼B户型样板房

220　Fuzhou East Great Town Show Flat
　　福州东方名城样板房

228　Show Flat of Sansheng Central Park
　　三盛中央公园样板房

238　Show Flat of Forte Ronchamps, Nanjing
　　南京复地朗香样板房

250　Hung Hom Lingshijun Pulin Garden — Showflat of Apartment A
　　红磡领世郡普霖花园A户型样板房

Chinese Style

中式风格

Hung Hom Lingshijun Pulin Garden—Show Flat of Apartment C

红磡领世郡普霖花园 C 户型样板房

Designer: Chen Yi, Zhang Muchen
Interior Design: Fenghe Muchen Space Design
Location: Jinnan District, Tianjin, China
Floor Area: 650 m²
Photography: Zhou Zhiyi

设 计 师：陈贻　张睦晨
室内设计：风合睦晨空间设计
项目地点：中国天津市津南区
建筑面积：650 m²
摄　　影：周之毅

Keywords / 关键词

New Chinese Style　新中式风格

Elegance　清雅幽远

Zen　禅意境界

Furnishings/Materials / 软装 / 材料

Solid Wood Flooring, Wallpaper, Stone Mosaic, Latex

实木地板、壁纸、石材马赛克、乳胶漆

Designers intend to present the implicited and ethereal zen in culture level with their unique understanding of Chinese culture. Modern way of moderling language is adopted in the space exploring the root of Chinese traditional culture; the comtinuition of cultural spirit and the presentation of qualified space are the most essential objectives of the project.

Removable wooden partitions adopted in the space come from the designers' understanding and master of Chinese traditional pattern which is the only traditional visual factors in the entire space. The application of partitions conveys the landsscaping theory in Chinese culture that matual generation between virtualness and reality, landscape intersection. Virtualness is presented through reality and reality is embelished through virtualness, forming the unique structure language way of the design; the entire spatial layout and atmosphere manisfest the taste of oriental life esthetics and wonderland.

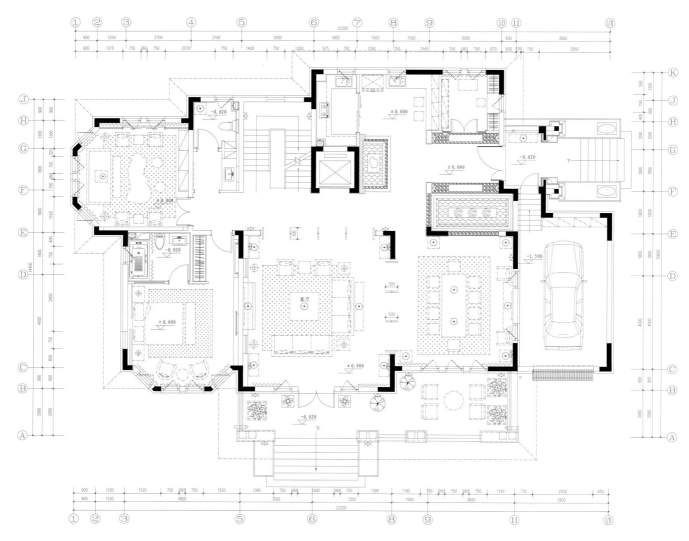

First Floor Plan
一层平面图

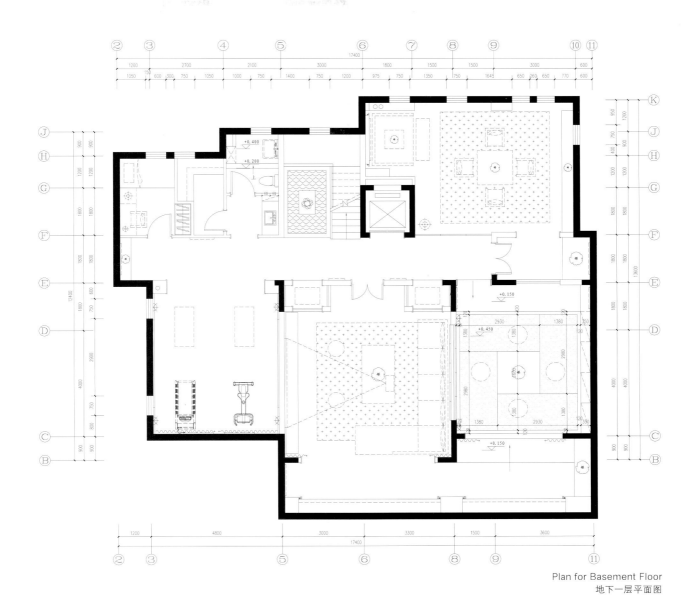

Plan for Basement Floor
地下一层平面图

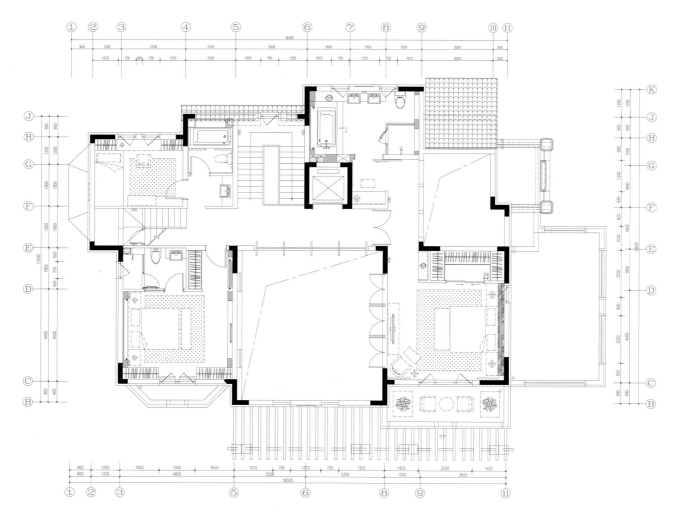

Second Floor Plan
二层平面图

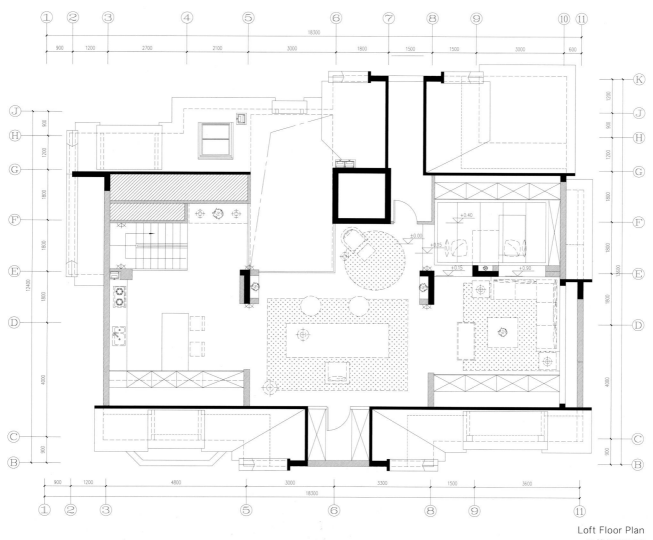

Loft Floor Plan
阁楼层平面图

设计师凭着对中式文化的独到理解，在此空间中更多推崇的是文化精神层面集含蓄和空灵为一体的禅意境界。空间中通过使用当代的造型语言方式去寻求中国传统文化脉络延续的根源；文化精神的延续和气质空间的呈现是此次设计寻求的最本质的诉求点。

该空间中大量运用了中国传统的木质构成的可移动隔屏，其形式内敛恬静，隔屏的形式来源于设计师对于传统纹样的理解和剖析，隔屏纹样几乎成为整个空间内唯一的传统视觉构成元素。隔屏的运用体现中国文化中讲究虚实相生景物相透的造景理论。虚境通过实境来实现，实境又在虚境的统摄下来渲染，虚实相生成为该设计独特的结构话语方式，而整个空间布局和氛围营造则呈现出东方生活美学心灵梦乡的格局。

MOON RIVER Show Flat 1

水月周庄样板房一

Chief Designer: Davies Xiao
Design Team: Zhan Yonglin, Cai Yajuan
Interior Design: X.S.Design
Location: Suzhou, Jiangsu, China
Area: 250 m²
Photography: Edward Xiao

设计主创：萧爱彬
设计团队：詹永林　蔡娅娟
室内设计：萧氏设计
项目地点：中国江苏省苏州市
面　　积：250 m²
摄　　影：萧爱华

Keywords / 关键词

New Oriental Style　新东方主义风格
Transparent & Generous　通透大气
Elegant Simplicity　朴素雅致

Furnishings/Materials / 软装 / 材料

Marble, Wood Facing, Floor, Glass, Emulsion Varnish

大理石、木饰面、地板、玻璃、乳胶漆

This is a new oriental style project that interprets both contemporary oriental life and times. Designers introduced oriental spirit into westernized space to present the unique life appearance of contemporary Chinese. Meanwhile, theme of lyre-playing, chess, calligraphy and painting was used to explore modern oriental lifestyle, producing a subtle, restrained yet romantic atmosphere that inherits and develops the tradition.

The original building space was optimized and replaced by a space with certain degree of relaxation and main color of elegant warm tone, in which the materials in same color system interweaved together, look contrastive and coordinated. Traditional decorative signs were arranged by modern design techniques to reflect modern oriental cultural deposit and get close to contemporary oriental aesthetic taste. The whole space design highlights the continuation of style and architecture, which provides an ideal living space for residents with simple and elegant decoration.

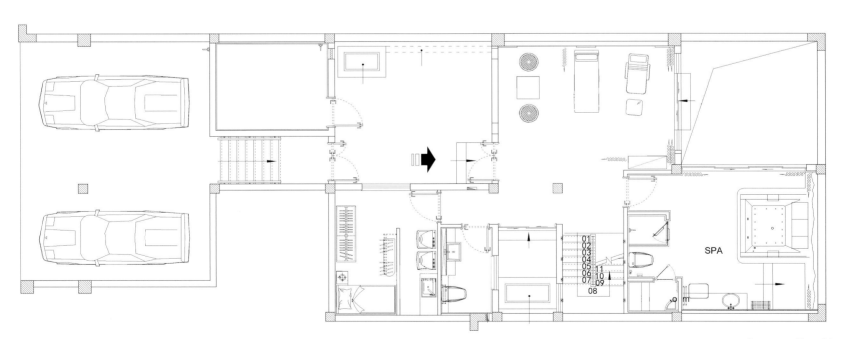

Basement Floor Plan
地下室平面图

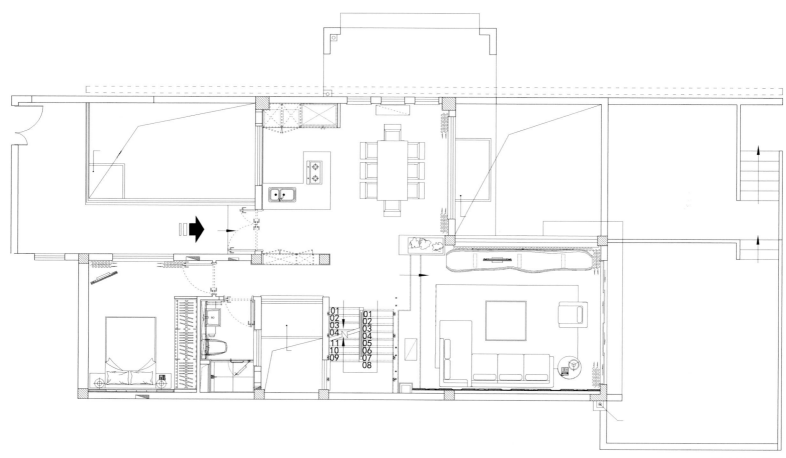

First Floor Plan
一层平面图

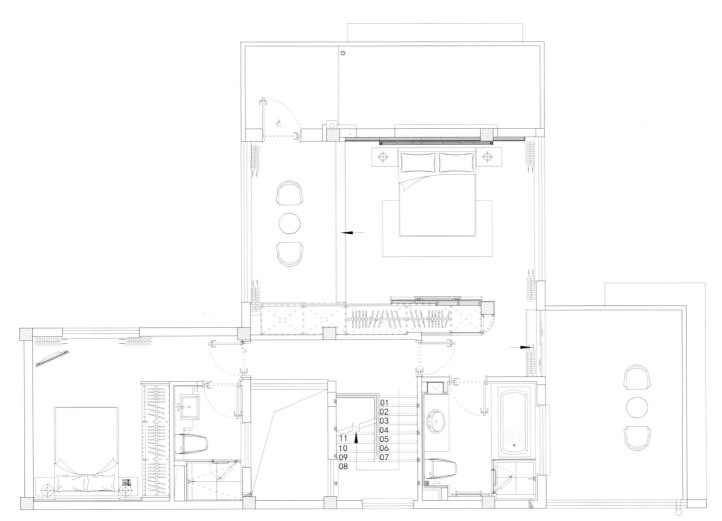

Second Floor Plan
二层平面图

　　本案为新东方主义风格，对当代东方生活形态与时代形式进行探索。设计过程中将东方精神注入过于西化的空间中进行思考，以呈现一种当代华人独有的生活样貌。本案又以琴棋书画为主题探讨现代东方的生活方式，创造既有传承又有发扬，既内敛含蓄又不失浪漫的意境。

　　本案首先将原有的建筑空间优化，务求让空间张弛有度，空间以素雅的暖色调为基调，同色系的材质相互穿插，既有对比又很协调。现代设计手法的灵活运用将某些传统的装饰符号重新铺排，让其体现现代东方的文化底蕴，也更贴近当代东方人的审美情趣。整个空间设计注重风格与建筑的延续性，装饰朴素雅致，构成文人居士理想的生活空间。

Show Flat of Oriental Charm, Changzhou

常州东方风韵样板房

Designer: Dong Long
Interior Design: DOLONG Design
Location: Changzhou, Jiangsu, China
Area: 500 m²
Photography: Jin Xiaowen Spatial Space Photography

设 计 师：董龙
室内设计：DOLONG设计
项目地点：中国江苏省常州市
面　　积：500 m²
摄　　影：金啸文空间摄影

Keywords 关键词	Chinese Style　中式风格 Zen Space　禅意空间 Environmental Protection Material　环保材料
Furnishings/ Materials 软装／材料	Wood Finish Tile, Terrazzo, Wood Carving, Pane, Wallpaper, etc. 木纹砖、水墨石、木雕、窗格、墙纸等

Standing from the needs and background of the owner, the designer applies Chinese traditional Zen elements to make appropriate orientation, planning and design, and forge a practical and environmental-friendly villa space, full of Zen.

In the design process, this designer conducts an in-depth exploration to the owner's residential requirement and value. The overall design is coordinated, beautiful and convenient for the owner's living. In environment style control, the designer creatively applies Chinese traditional elements partially in modern space to highlight the essence of Chinese culture. In spatial layout, the project emphasizes on breakthrough of ideas and the unity of spirit. In material selection, the project promotes environment-friendly wood furniture and materials to receive a better effect, which is also in consistent with the owner's requirement.

　　本案从别墅主人的需求与背景出发，运用中国传统文化中的禅意元素去合理地定位、规划、设计，打造了一个体现禅意、实用环保的别墅空间。

　　在设计过程中，设计师对业主的居住需求、生活价值等进行了深入挖掘，整体设计协调、美观，同时也使得业主居住起来很方便。在环境风格的把握上，其设计创新点在于将现代的空间和局部点缀中式元素来体现禅意的空间，彰显出中式文化的精髓；在空间布局上注重突破思想，围绕一个中心，使得项目呈现一个完整的整体；在设计选材上提倡环保，选用原木家具、环保材料，从而创造了良好的环保、实用效果，并且契合了业主对于项目的要求。

Show Flat of Shenzhen Jixin City

深圳集信名城样板房

Designer: Wang Wuping
Interior Design: Shenzhen Taihe South Architectural & Design Design Office
Developer: Shenzhen Jixin Investment Development Co., Ltd.
Location: Shenzhen, Guangdong, China
Area: 300 m²

设 计 师：王五平
室内设计：深圳太合南方建筑室内设计事务所
开 发 商：深圳市集信投资发展有限公司
项目地点：中国广东省深圳市
面　　积：300 m²

Keywords / 关键词

New Chinese Style　新中式风格

Cultural Deposits　文化底蕴

Unique and Elegant　独特高雅

Furnishings/Materials / 软装 / 材料

Fraxinus Mandshurica Brush off Paint, Wallpaper, Glazed Ceramic Tile, Mushroom Slate, Mosaic, etc.

水曲柳擦色漆、墙纸、抛釉砖、蘑菇石、马赛克等

This project is located in Shajiang East Road, Songgang, Shenzhen with mountains around and ecological beauty of the mountain park, which is an ideal living environment for the upper class. It has unique natural scenery, 300 m² open visual field and large hollow home gardens, creating a new Chinese style visual feast being unique and elegant, steady and grand and rich in cultural deposits.

The adoption of white pebbles, rough mushroom slate, vivid and interesting waterscape, original outdoor floor and two chic benches displays the essence of the garden everywhere. With the chic home garden, surely, the interior space intends to share its beauty. On the facade renovation, the brick wall between the home garden and sitting room is broken through and designed into a Chinese partition screen, so people can sit in the living room and enjoy the environment.

The wall between the dining room and corridor is designed a hollow screen bearing the same elements as the home garden, which increases lighting condition of the corridor and visually brightens the dining room

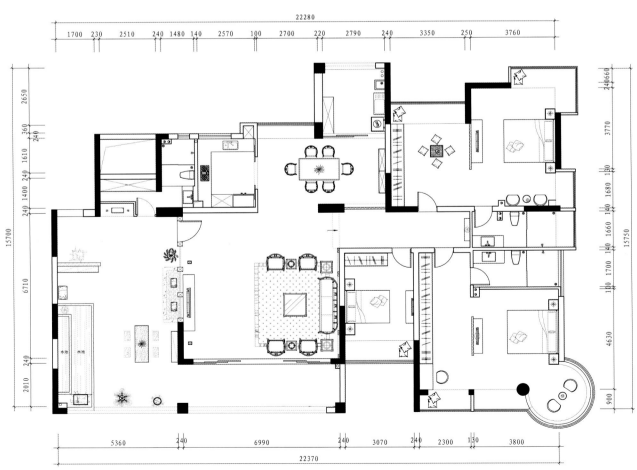

Floor Plan
平面布置图

as well. Walls on both sides of the kitchen apply fraxinus mandshurica brush off paint, high-grade and steady, which brings strong contrast effect in the visual space of the sitting room. The kitchen door also uses the same screen elements without door pocket, integrated with the wall as a whole; with the artistic sense of the screen partition, it plays down the concept of the door.

The room adopts some simple techniques, but not loses the Chinese elements. The TV partition uses Chinese hollow flowery decoration, and the bedside uses wallpapers shaped which is fashionable and delicate but not loses Chinese charm.

本项目位于深圳松岗的后花园沙江东路，群山环抱，尊享山体公园生态美景，集山水天然优势，使其成为上流社会生活的领地。该样板房有着得天独厚的自然美景，300 m² 的大开度视觉享受，超高大的中空入户花园，打造出一场独特高雅、沉稳大气、有文化底蕴的新中式视觉盛宴。

白色的鹅卵石，粗犷的蘑菇石材，生动的水景，原汁原味的户外地板，加上两条别致的石凳，无处不在演绎园林的精粹。有着这样大气别致的入户花园，室内空间自然也要分享其美。在立面改造上，把入户花园和客厅之间的砖墙打通，设计成带有一点中式情结的隔断屏风，这样坐在客厅就能静观其境。

在餐厅与过道中间，将一面墙打通设计成与入户花园元素相同的镂空屏风，增加过道的采光与视觉拉伸，同时也为餐厅增色不少。厨房的两面墙全设计成

水曲柳擦色工艺漆面,高档沉稳,在客厅大视觉空间里,起到强烈的对比作用。厨房的门也采用同样的屏风元素,没有门套,和墙面浑然一体,有着屏风隔断的艺术感,而淡化了门的概念。

房间则采用一些简洁的手法,但不失中式情结的元素。电视隔断采用中式镂空隔花,床头背景运用"回"字型的墙纸饰面,时尚精致而又不失中式余韵。

Show Flat of Beijing Yanxitai No. 37

北京燕西台 37 号样板房

Designer: A Li, Wang Boyu, Liang Hao
Interior Design: Ideal Design and Construction Co., Ltd.
Location: Haidian District, Beijing, China
Area: 400 m²

设 计 师：阿栗 王博宇 梁浩
室内设计：北京艾迪尔建筑装饰工程有限公司
项目地点：中国北京市海淀区
面　　积：400 m²

Keywords 关键词

Modern Chinese Style　现代中式风格
Cultural Connotation　文化内涵
Convenient and Comfortable　方便舒适

Furnishings/Materials 软装/材料

Pendant Lamp, Wood, Floor, Carpet, Decorative Painting

吊灯、木材、地板、地毯、装饰画

This project aims at combining traditional and the modern elements. The traditional classical design brings peace to the people living in the city, while the modern elements meet the demands for modern life and present the fashionable taste of modern time. The design for the ground floor courtyard integrates the Chinese traditional culture elements in the bridges and the pavilions like the Jiangnan Watertown or the Jiangnan Gardens.

The functional layout meets not only the youth's demand for amusement, but also the elder's demand for quiet, convenient and comfortable living space.

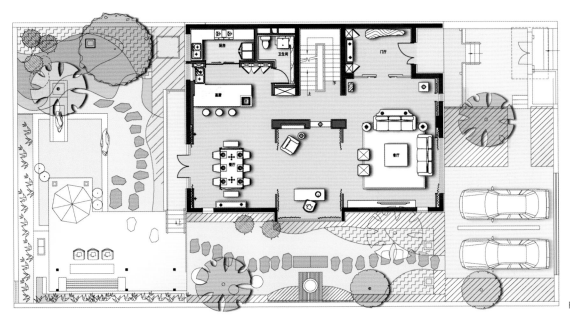

First Floor Plan
首层平面图

该样板房设计旨在将传统与现代紧密结合，传统古典的设计使人在浮华的都市中得到片刻的沉静，而设计中现代元素的运用不仅满足现代生活的需求，同时体现了对于现代时尚的品味。首层的庭院设计充分融入了中国的传统文化元素，小桥流水、山水亭台，漫步其中，仿佛身处江南水乡或是江南园林之中。

在功能布局上，既考虑到青年人对时尚、娱乐、休闲的需求，同时又考虑到长辈对安静、方便舒适的生活空间的需求。

Type A1 Show Flat of Junyue Golden Beach, Zhangzhou

漳州君悦黄金海岸A1户型样板房

Chief Designer: Qiu Chunrui
Interior Design: Shenzhen Dayi Interior Design Ltd.
Developer: Zhangzhou Hongye Tongchuang Real Estate Co., Ltd.
Location: Zhangzhou, Fujian, China
Area: 68 m²

主设计师：邱春瑞
室内设计：深圳大易室内设计有限公司
开 发 商：漳州鸿业同创房地产有限公司
项目地点：中国福建省漳州市
面 积：68 m²

Keywords / 关键词

Chinese Style 中式风格

Zen Space 禅意空间

Elegance 典雅脱俗

Furnishings/Materials / 软装/材料

Volakas White, Beige Travertine, Perlino Bianco, Silver Mirror, Wall paper, Wood Floor, Leather

爵士白、米黄洞石、白木纹、银镜、墙纸、木地板、皮革

This project is like an anchorite who is far from the bustle world and enjoying the solitude.

The artistic conception of the wash painting is smashed and applied in the space, while the simple and straight lines are used to display the elegance of the neo-orientalism. Some furniture with traditional elements are put in the simplified Chinese space to emphasize the oriental charm in the room. Whether the stylish containers and the lamps, or the abstract splash-ink paintings, or the grilling with Zen style, they all present the classic flight with the invisible language. The warm light leads the space to the classical and mysterious conception and makes visitors linger.

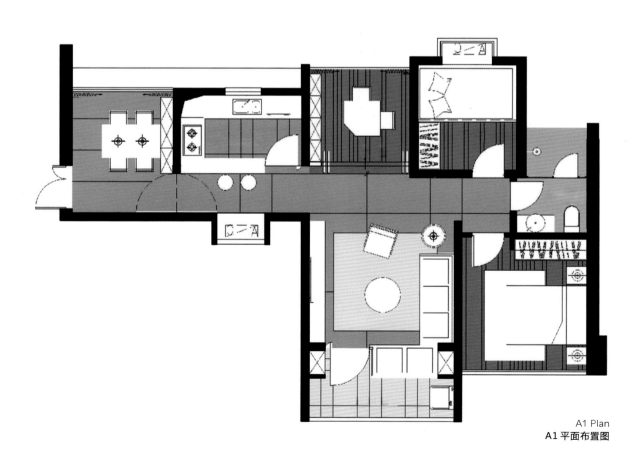

A1 Plan
A1 平面布置图

　　本案犹如一个隐者,远离尘世,独享着偏安一隅的惬意。

　　水墨画的写意意境被设计师打碎后渗透到这个空间中,并且用众多简洁平直的线条加以演绎,表现出新东方主义的清雅大方。在简化的中式空间里,设计师特意放入了一些传统符号作为明显的陈设,以加强室内典雅脱俗的东方韵味。无论是造型考究的器皿、灯具,还是画面抽象的泼墨画,亦或是传递出淡淡禅意的格栅……无不体现出一种无法言喻的复古情怀。温和的光源,将空间引入到一种古典且神秘的意境之中,令置身其中的人不禁流连于此。

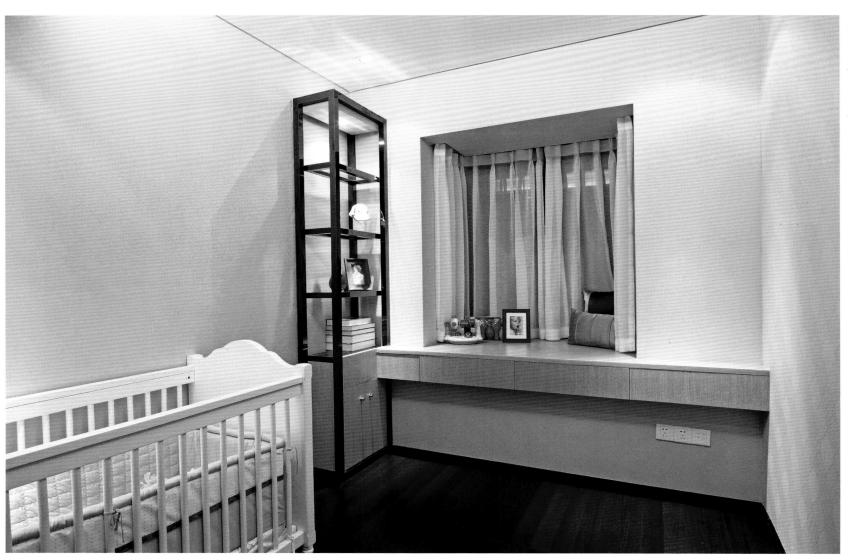

Top Inkstone Residence, Taoyuan

桃园一品国砚

Chief Designer: Zhang Xin
Interior Design: CYNTHIA'S Interior Design Studio
Location: Taiwan, China
Area: 198 m²

主设计师：张馨
室内设计：张馨/瀚观室内装修设计工程有限公司
项目地点：中国台湾
面　　积：198 m²

Keywords 关键词

Chinese Style　中式风格
Classical Sense　古典气息
Connected Spaces　空间通透

Furnishings/Materials 软装/材料

Window Grille, Wooden Floor, Painting
窗花、木地板、挂画

In the living room, there is a screen designed with Suzhou window grille which is an important element in traditional Chinese style. In addition, Suzhou is the hometown of the owner, and the design will remind the owner of his precious memory. The floor tiles of the passage and the living room have changed in size and color to define the layering of the space. The TV wall is designed with dark and light colored wood as well as understated lines, to give the nostalgic taste together with the customized TV stand. The black and white lighting fixtures made of glossy material reflect a high-quality living space. All the furnitures are simple and elegant with exquisite details. For example, the sofa having simple lines and dark-colored edges, feels delicate and classical.

A flower stand is put between the dinning room and living room as a partition. In the nostalgic dinning room, there are rattan chairs and a dark-colored wooden table to remind the owner of his old house. The ower is in love with nature, therefore, the designer puts a painting of forest on the wall.

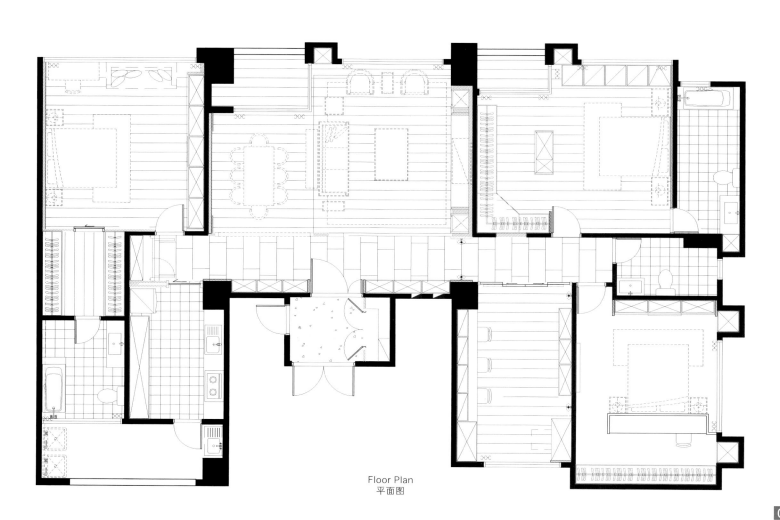

Floor Plan
平面图

In the master bedroom, there is a large-sized bed for the couple to read and enjoy views sitting beside the window. The dressing area is designed with beige veneer which is decorated with classical lines. Since the elder brother of this family is preparing for an important examination and he needs to study hard and long, the designer adds a small side table for dishes and snacks. While the bedroom of the young brother is decorated in white and lake blue with a desk at the bedside for partition.

客厅挑选来自苏州的窗花作为屏风，是中国复古风格不可或缺的重要元素，正巧苏州也是业主的家乡，可以使业主产生亲切感。地砖从廊道跨越至主空间的变化，有从小变大、由暗到明的层次感。电视墙采用颜色一深一浅的木作刻画中式的内敛线条，搭配设计师绘图订制的电视柜，在怀旧味道中又增加了精致感。客厅内黑白两色的亮面材质的原装灯具，衬托出居家应有的质感。在中式风情的空间中，设计师不完全采用极简人文家具，如沙发虽然线条简单，却因深色勾边的小细节，增添细腻的古典味。

餐厅望向客厅的部分透过端景柜划分前后两区，充满怀旧古朴气氛的餐厅，选以藤编椅、深色木桌，不经意间业主就有老房子里的记忆。屋主十分向往大自然，设计师在餐厅的背景墙上挂了一幅林间图，空间自然气息浓厚。

主卧室里设置了特别加宽的卧榻，业主夫妇二人便能坐卧在窗旁阅读、欣赏风景。更衣区饰以古典线条的门片，淡淡的黄色木皮复古情味浓厚。哥哥处于准备考试的关键时期，需要长时间在书桌上苦读，因此设计师在哥哥房内增加小边几，让家人能随时放茶点给哥哥补充体力。弟弟房以白、湖水绿为主视觉，方正的房间更重视机能配比，故利用书桌作为床头及隔间。

Show Flat of Lonyear Villa Experience Hall

苏州朗誉别墅装饰实景体验馆样板间

Designer: Wu Junyi
Interior Design: Lonyear House Decoration Engineering Co., Ltd
Location: Suzhou, Jiangsu, China
Area: 200 m²

设 计 师：巫俊逸
室内设计：朗誉别墅装饰工程有限公司
项目地点：中国江苏省苏州市
面　　积：200 m²

Keywords / 关键词

New Chinese Style　新中式风格
Simplicity & Purity　简约纯净
Warmth & Elegance　温馨优雅

Furnishings/Materials / 软装 / 材料

Marble, Oak, Floor, Mosaic, Ceramic Tile

大理石、橡木、地板、马赛克、瓷砖

The project attends to express a elegant and reserved, dignified and luxurious oriental atmosphere through penetration of Chinese elements in space, and the classicism in peaceful decoration, providing a new sensory experience. The design has implanted the soul of new era as well as the classic and immortal Chinese ancient patterns.

The TV background wall in the living room made of Volakis marble highlights a simple design style while the horizontal and vertical black wooden frame brings the contrast between the structure and colors. White set of porcelains and celadon are positioned asymmetrically around the decorative panting, raising the spatial charm. The green orchid brings the vitality of life to the space, responding to the carps in the pond.

The mosaic stickers in the open kitchen and the silver mirror on the wall of dining room features modernization, and space is stretched through the silver mirror ceiling, creating a visual sense of dimensional dislocation. The kitchen in simple design enables the owners' better enjoyment of cooking time. Chinese ginghams are adopted to collocate with the dinning table and the lotus pond provides a enjoyable

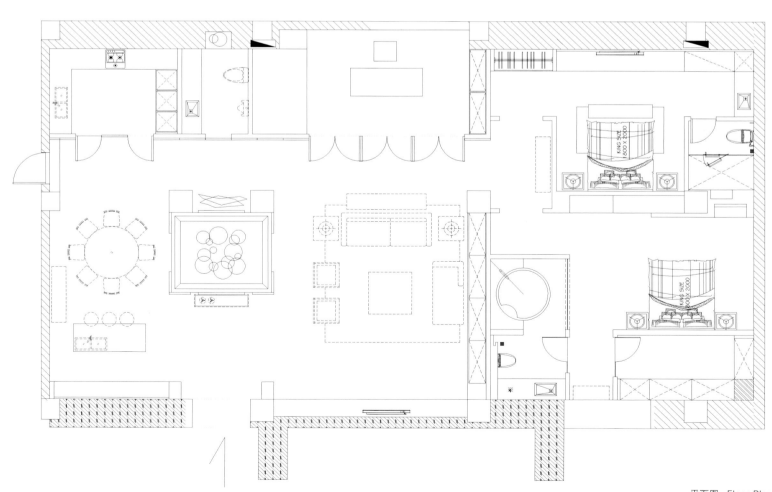

平面图 Floor Plan

dining environment to the diners.

The red decorative painting in the bedroom entrance leads the vision line inside the bedrooms. The tranquil and graceful main bedroom occupies furniture featuring oriental colors with simplified lines on the basis of traditional Chinese furniture, in pursuit of reserved and unadorned style and practicable space. The mosaic and silver mirror put the whole bedroom in harmony and unity, with the embellishment of pure white sanitary wares strengthening the modern sense and extending the general style of simplicity and purity.

The secondary bedroom provides the comfortable and warm sense through the handmade wool carpet. The atmosphere extends to the wall through a large area of wallpapers and the wooden lines and bed extend the spatial structure. The lighting and wallpapers in simple but elegant colors creates a calm and comfortable atmosphere and environment for rest and sleep. The ornaments in alcove carry on the general spatial conception and become one of the visual focuses.

本案通过中式元素在空间中的渗透，平静的装饰中蕴藏经典的中式韵味，给人一种全新的感官体验，向人们传递清雅含蓄、端庄丰华的东方式精神境界。整个设计在传承中国古老纹样的经典与不朽的同时又注入了新时代的灵魂。

客厅电视背景的整面爵士白大理石墙没有过多的造型，强调了简洁的设计风格，一横一纵的黑色装饰木架增加了结构和颜色的对比。白色组合瓷器和青瓷以装饰画为中心非对称的摆放，增加了空间情调，绿色的兰花给空间带来了生命的活力，与水池的鲤鱼形成呼应。

开放式厨房马赛克贴饰与餐厅墙面的银镜贴饰尽显现代化，吊顶的银镜更是拉伸了空间，给人立体错位的视觉感受，简约的厨房让业主更好的享受烹饪时间，餐桌边上搭配中式花格，水池荷花为就餐者提供了非常惬意的用餐环境。

卧室玄关红色装饰画把视线引入卧室，设计师为主卧营造出了恬静和优雅的居室氛围，家具在传统中式家私的基础上把线条简化，但又不失东方色彩，更迎合追求内敛、质朴的设计风格，使空间更加实用。马赛克和银镜使整个卧室显得和墙谐而统一，点缀纯白色的卫浴，不露痕迹地加深了现代感，此处延续了简约、纯净的整体风格。

次卧舒适而温暖的质感由手工羊毛地毯倾情营造，大面积的壁纸把这样的气息一直延伸到墙面，木饰线条与床延伸了空间的结构。灯光和壁纸的颜色虽然素雅，但它提供了一个平和安逸的休息氛围和睡眠环境。壁龛中式的摆件延续了整个空间的意境，成为空间的一个视觉焦点。

Mediterranean Style

地中海风格

Type A Show Flat of Red Coral Bay, Shijiazhuang

石家庄红珊湾 A 户型样板房

Designer: Wang Weidong
Interior Design: D.H. Top Interior Design Office
Location: Shijiazhuang, Hebei, China
Area: 150 m²

设 计 师：王卫东
室内设计：东合高端室内设计工作室
项目地点：中国河北省石家庄市
面　　积：150 m²

Keywords / 关键词

Spanish Style　西班牙风格

Noble Temperament　贵族气质

Living Ambiance　居家氛围

Furnishings/Materials / 软装 / 材料

Vintage Wooden Flooring, Tiles, Wallboard, Diatomaceous Earth, Painting

复古的木地板、瓷砖、墙板、硅藻土、挂画

It is a 150 m² unit with three bedrooms. Spanish-style design will make people feel warm and comfortable. Luxurious and elegant living room, pleasant and comfortable bedrooms as well as the multi-level space structure will bring romance and relaxation to life. With vintage wooden tiles and ceramic tiles on the floor, and the white wallboards and diatomaceous earth on the wall, the space will look natural without the sense of monotonousness. Yellow color of the diatomaceous earth reminds people of the warm sunshine, while the paintings on the wall tell the story of the space.

The wooden ceiling matches the whole design, and the traditional hand-painting shows multi-elements in Spanish style. Design techniques here integrate with art perfectly. Flexible iron decorations become joints between different spaces.

本套示范单元为150 m²的三居室。西班牙风格的家居最容易营造出温暖舒适的居家氛围，质朴中透出贵族气质的客厅、舒适温暖的卧室，多重立体的空间结构让生活多了些浪漫和随意。复古的木地板和瓷砖的结合让地面不再枯燥乏味，墙面以恰当的比例结构合理配置了做旧的白色墙板和硅藻土，使墙面看起来不再那么单调，硅藻土的天然质感自然亲切，暖黄色像阳光一样，让视觉和触觉同步。墙面错落有致的挂画，伴着满室阳光表达着空间的故事。

顶面温润的木质结构把整个顶面和谐地统一在一起，手绘的传统图案让西班牙风格的多元化体现得淋漓尽致，技术与艺术在这里完美融合。灵动的铁艺以艺术的名义让空间与空间之间变得更加完美。

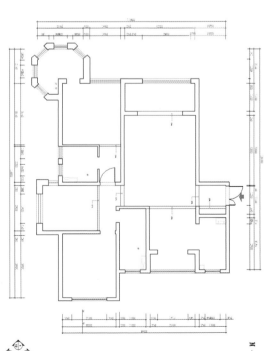

平面图

Neo-classical Style

新古典主义风格

Show Flat of Fuzhou Sansheng Central Park

福州三盛中央公园样板房

Designer: Zhu Wenli
Plan Approver: Ye Bin
Interior Design: Fujian GuoGuang YiYe Architectural Decoration Design Engineering Co., Ltd.
Location: Fuzhou, Fujian, China
Area: 170 m²

设 计 师：朱文力
方案审定：叶斌
室内设计：福建国广一叶建筑装饰设计工程有限公司
项目地点：中国福建省福州市
面　　积：170 m²

Keywords 关键词

Neo-classical Style　新古典主义风格

Concise Line　简洁线条

Multi-Level　多层次感

Furnishings/Materials 软装 / 材料

Through-Body Tile, Real Wood Floor, Elegance Beige Marble, Wallpaper, Sprayed Tawny Glasses, Soft Leather Bag

通体砖、实木地板、圣雅米黄大理石、壁纸、喷花茶镜、皮革软包

This project mainly applies the neo-classical style with the characteristics of being elegant, delicate and its rich connotation. Like the diversified thinking model, it combines the romantic meditate feelings and modern people's demand for life, and blends luxury, elegance, vogue and modern together to reflect the personalized aesthetic views and cultural taste of the post-industrial era.

Based on the original reasonable and perfect space structure, the space layout applies concise and fluent lines to outline the space, letting it get unlimited extension on the plane and abound its sense of space levels on the facade, thus adding more lively space looks. In materials the integrated use of all kinds of marbles, forsted glass, mirror surfaces, painting boards, art mild packages, wallpapers and wood makes different spaces bear different looks and temperaments; and finally these all go around a theme — the quiet and elegant style.

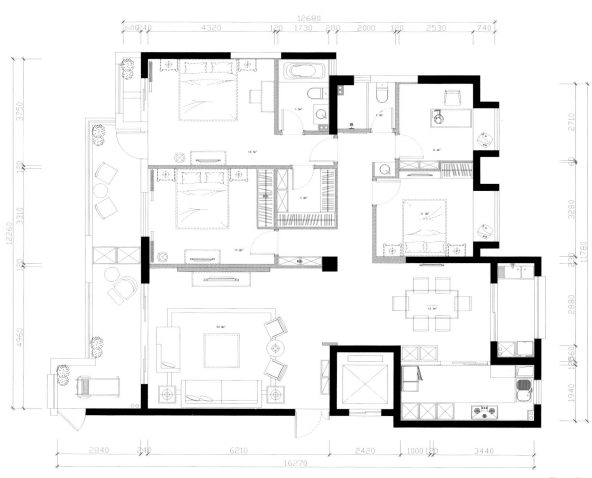

Floor Plan
平面图

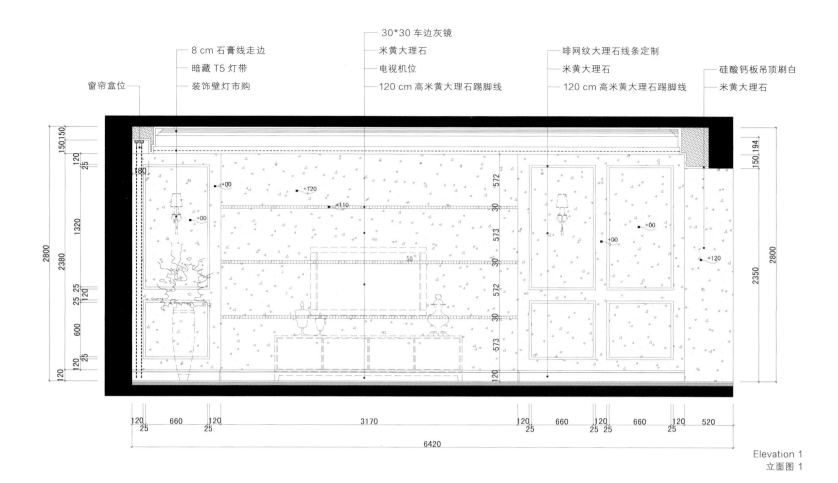

Elevation 1
立面图 1

本案以新古典主义风格为主，优雅、精致而富有内涵是新古典主义风格的特点。像是一种多元化的思考方式，将怀古的浪漫情怀与现代人对生活的需求相结合，兼容华贵典雅与时尚现代，反映出后工业时代个性化的美学观点和文化品位。

本案空间布局上，原本的空间结构合理而完善，在此前提下，以简洁流畅的线条勾勒空间，让空间在平面上得到无限延伸，在立面上丰富了空间层次感，让空间表情更为生动。在材料运用方面，各种大理石、喷花玻璃、镜面、烤漆板、艺术软包、墙纸、实木等综合运用，让不同的空间拥有不一样的表情和气质；而这些，最终都围绕一个主题——低调而典雅的风格。

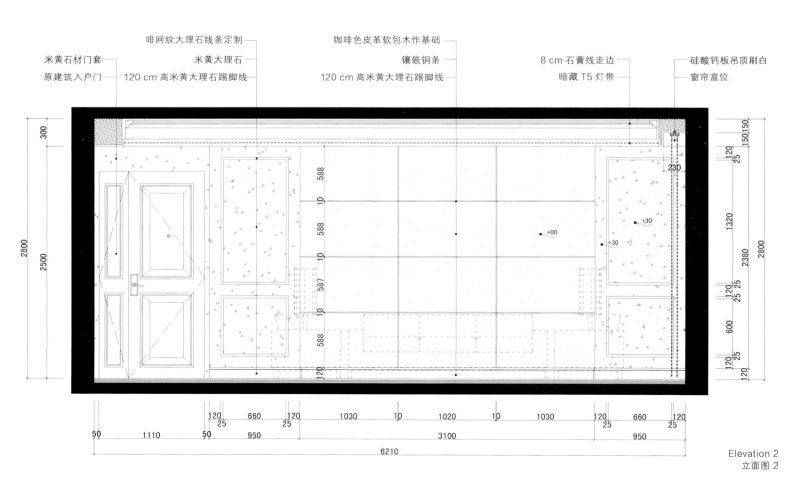

Elevation 2
立面图 2

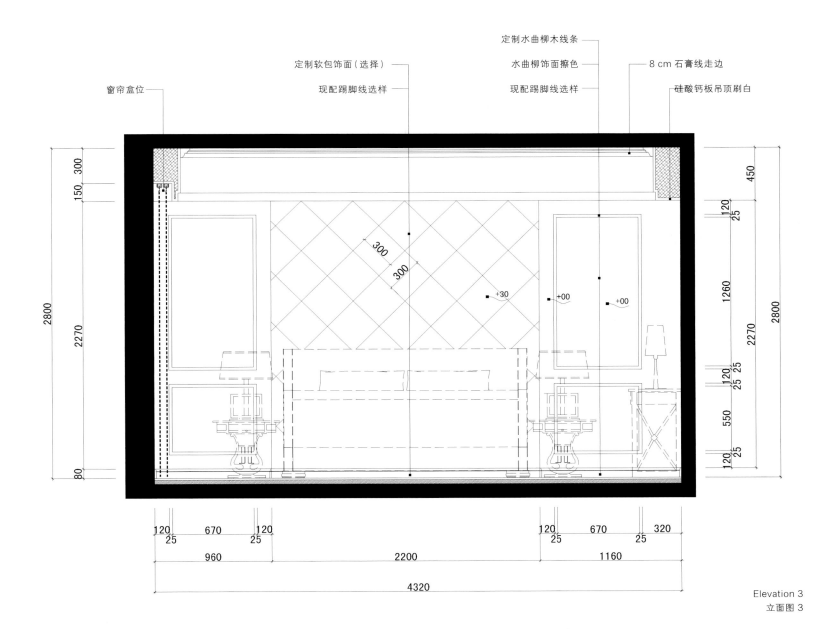

Elevation 3
立面图 3

Show Flat of East Shore International Garden, Hohhot

呼和浩特东岸国际花园样板房

Designer: Zhu Jianshen
Interior Design: Depth Interior Design Office
Location: Hohhot, Inner Mongolia, China
Area: 360 m²

设 计 师：祝建深
室内设计：深度室内设计事务所
项目地点：中国内蒙古自治区呼和浩特市
面　　积：360 m²

Keywords 关键词

Classical Style　古典风格
False or True Complement　虚实相生
Transparent Space　空间通透

Furnishings/Materials 软装/材料

Gardenia Tile, Milosi Marble, Nature Board

加德尼亚瓷砖、米洛西大理石、大自然地板

The designer has been deeply communicated with the owner before the design of this 360 m² duplex house. The owner prefers more modern and bright feelings on the basis of the classical style, so in the plane planning, the designer tries to bring in more natural light inside as far as possible to make the original open space more transparent. In the shape making, the designer eliminates the Europe type style of heavy and complicated feelings; for the color tone, it adopts bright champagne and grey color to reflect the transparence of the space. The collocation of modern design elements and classic furniture shows another kind of unique temperament. The adoption of harmony transitional elements in the whole space makes the space in false or true complement with open visual field; in addition, the unified colors and exquisite materials fully reflect the low-key luxury of the space.

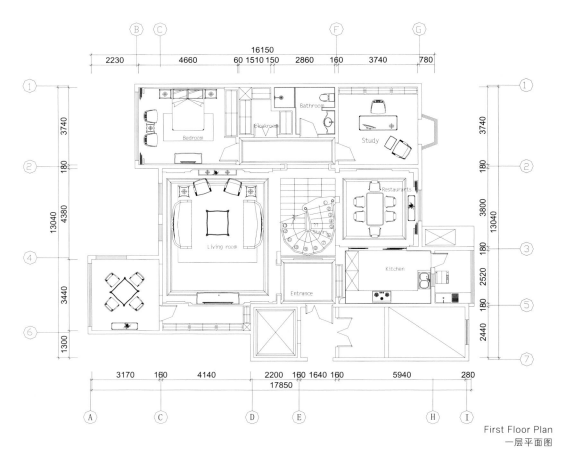

First Floor Plan
一层平面图

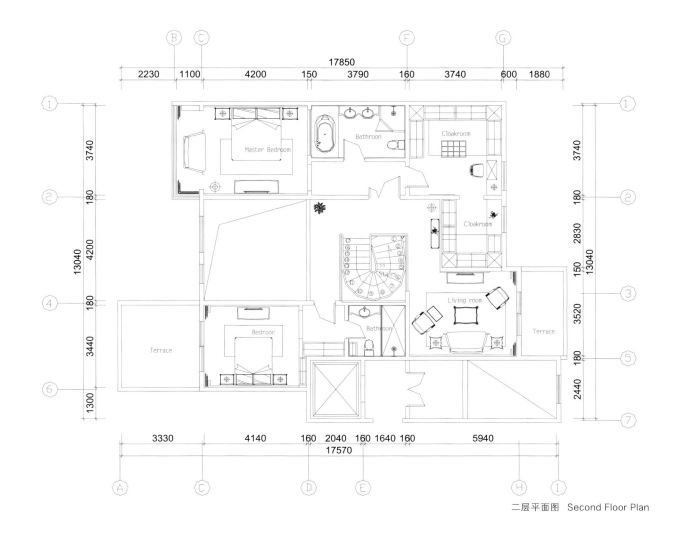

二层平面图 Second Floor Plan

本案是一套360 m²的复式结构住宅，在做设计前与业主进行了深层沟通。业主要求在古典风格上多体现一点现代明亮的感觉，所以在平面规划的时候尽可能地把自然光线引进室内，使原本比较开敞的空间更具通透感。在造型上，省去了欧式风格的繁复感；在色调上，尽量以明快的香槟色以及灰白色调去体现空间的通透感。用现代设计元素搭配古典家具，所表现出来的是另一种独特气质。在整个空间中运用和谐的过渡元素，使空间虚实相生，视野开阔，加上色彩的统一、用材的讲究，尽显低调的奢华。

Type B Show Flat of Coral Bay, Shijiazhuang

石家庄红珊湾 B 户型样板房

Designer: Niu Guohua
Interior Design: D.H. Top Interior Design
Location: Shijiazhuang, Hebei, China
Area: 130 m²

设 计 师：牛国华
室内设计：东合高端室内设计工作室
项目地点：中国河北省石家庄市
面　　积：130 m²

Keywords 关键词	Neo-classical Style　新古典主义风格 Peaceful and Graceful　宁静优雅 Fashionable and Dynamic　时尚动感
Furnishings/ Materials 软装/材料	Fabric Sofa, Stone, Capet, Crystal Chandelier, Painting, Floor, etc. 布艺沙发、石材、地毯、水晶吊灯、挂画、地板等

For people who love fashion and look for a convenient working environment, the small houses in downtown area are the best choices for them. However, with this unsatisfactory layout, the designer should keep the idea of sustainable development in mind and use professional design techniques to create a fashionable and flexible living space.

Before design, the designer made an analysis on the target clients and then defined the theme of "dynamic, sharing and fashionable". The design starts from the hallway and then extends to the living room, master bedroom and other spaces to create a rational and personalized living atmosphere.

When opening the door, white colors of different levels will catch the eyes. Living room and dining room thus look quite elegant and graceful. White fabric sofas and white walls match with the graceful brown to shape a three-dimensional space. The design interprets neo-classical style with modern techniques. Lines of Louis XIV style are reserved and the redundant decorations are abandoned.

对于追求生活时尚和工作便利的都市潮人来说,市中心的小户型往往是他们的首选,但是面对有诸多不足的房型,要营造充满时尚且能灵活应对未来和功能兼备的家居空间,设计师就要以可持续发展的眼光、专业的设计手法去完美诠释和精心打造了。

设计之初,设计师对目标客户进行全面分析后,围绕着动感、共享、时尚的元素定位,从门厅着手,并以此主

线去整合与客厅、主卧以及各个空间的功能关系,来营造一个既理性又有个性的居住氛围。

打开房门,各种色阶的白色展现在眼前,白色调使客厅、餐厅显得非常宁静、优雅。白色搭配得非常巧妙,从白色布艺的沙发到白色造型的墙体,这些不同的白色材质配以淡雅的咖色,让整个空间立体了起来。设计师用现代的设计手法来诠释新古典的韵味,保留了路易十四的线条,去除了过多的繁琐装饰。

American Style

美式风格

Show Flat of Lake and Villa, Harbin

哈尔滨湖与墅样板房

Hard Decoration Designer: Hu Qiang
Soft Decoration Designer: Hao Yaxiao
Design Consultants: Gu Xiangfei, Zhao Dan
Interior Design: Beijing Fashion Impression Decoration Co., Ltd.
Furnishing Design: SANCY
Location: Harbin, Heilongjiang, China

硬装设计师：胡强
软装设计师：郝亚筱
设计顾问：古翔飞　赵丹
室内设计：北京风尚印象装饰有限责任公司
陈设设计：尚饰界
项目地点：中国黑龙江省哈尔滨市

Keywords / 关键词

American Style　美式风格
Art Sense　艺术气息
Multi-level　层次丰富

Furnishings/Materials / 软装/材料

Spanish Cream-colored Stone, Decorative Painting, Crystal Chandelier, Stone
西班牙米黄石材、装饰画、水晶吊灯、石材

American style satisfies people with simple and clear lines as well as elegant decorations. It features the simplicity, elegance and luxury in both modern style and European style. In this show flat, carvings and art wallpaper are used to highlight this style. And it retains the classical form and avoids redundant elements. Under the influence of modern aesthetics, it pays more attention to the functions and harmony of the decorations.

The hallway is well designed to adjust the layout and show the quality and style of the villa. At the same time, it enriches the spaces and well separates the interior form the exterior. With colorful stone patterns, unique ceiling and wallpaper, it highlights the North American style and interprets an elegant lifestyle.

In the living room, exquisitely designed fireplace and TV cabinet are not only functional but also decorative. While the dining room is designed with decorative lines to be another lightspot of this design.

The design for the study is very important. Delicate white lines are used to highlight is American style. The geometric division of the ceiling and the classics division of the walls as well as the art paint on the wallpaper integrate the whole space together. In addition with some motley-colored furniture, it well interprets the unique historical and cultural connotation in North American Style.

Bedroom is designed with colors of the same scheme. Pure background matches colorful furniture, decorations and fabrics to create an integral space.

American-style wallpaper in the staircase and the wooden material of the same color well define its luxury style. Due to limited area, the design emphasizes on "small but exquisite" to satisfy the villa life. The key selling point of this villa is the connection with nature. It is exquisite, elegant, low-key and luxurious at the same time.

美式风格，以其简洁、明晰的线条和得体有度的装饰，满足了人们对于现代风格与欧式风格的简洁、优雅、奢侈的需求。本案设计时，用雕花、艺术壁纸充分展现美式风格的艺术气息。在形式上保留了古典主义中的经典造型，并且不过于繁复。在现代审美的影响下，造型上更注重装饰的实用性和色调的和谐性。

改造后的门厅加了玄关，不但化解了风水上客户忌讳的格局，而且使其更有利于表现别墅的气势与品质，同时也丰富了空间及参观动线的变化，很巧妙地区分了空间的"内"与"外"。通过丰富的石材拼花、独特的造型顶面及壁纸等设计手法来表现北美风格，以展现主人优雅的生活情调。

客厅里带有丰富造型和线条层次感的体量壁炉和视听柜，既满足了功能需求也起到了装饰墙面的作用。餐厅古典的分割形式搭配丰富的装饰线条是本案中重要的表现手法之一。

书房的设计占据着非常重要的地位，本案中书房采用造型细腻、层次丰富的白色线条来诠释美式的内涵。除顶面几何分割及墙面古典分割的造型外又覆盖了一层来自墙面壁纸底色的艺术漆，使得整个空间浑然一体，搭配色彩斑驳的合适体量家具，把北美风格特有的历史沉淀感及文化底蕴表现得淋漓尽致。

卧室丰富的造型及变化统一在同一色系内，干净的底色搭配色彩变化丰富的家具、饰品与布艺，实现了软、硬装两者之间的对立统一，使得房间整体完备但不失细节。

楼梯间美式壁纸加上同一色系内丰富的木质材料，将特有的低调奢华气质表现到了极致。空间的原始体量比较小，因此整个方案的调整也是让其小而精致，并不失去别墅生活的硬件装饰的要求。其最大的卖点是与自然相融，精巧不失大气、低调不失奢华。

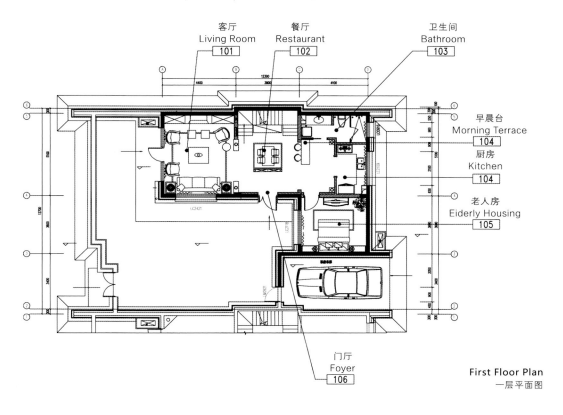

First Floor Plan
一层平面图

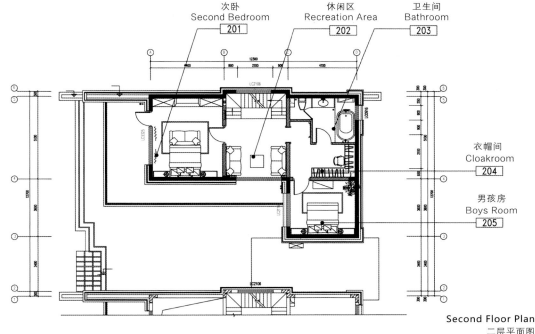

Second Floor Plan
二层平面图

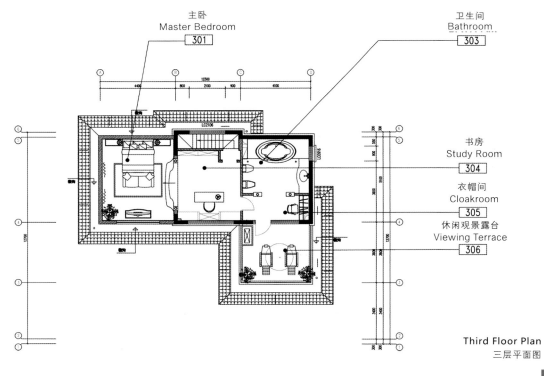

Third Floor Plan
三层平面图

Huangshang Paradise City House Type C-2 Show Flat

黄山和泰国际城 C-2 户型样板房

Designer: Liang Suhang
Location: Huangshan, Anhui, China
Area: 139 m²

设 计 师：梁苏杭
项目地点：中国安徽省黄山市
面　　积：139 m²

Keywords 关键词	American Country Style　美式乡村风格 Fresh and Elegant　清新淡雅 Harmonious Colors　色调和谐
Furnishings/ Materials 软装 / 材料	Wallpaper, Carpet, Throw Pillow, Pendant Lamp 壁纸、地毯、抱枕、吊灯

Color relationship is the first thing in living space design, no matter what the color scheme is, designers should be sure that a harmonious effect is obtained. Wallpaper, ceiling and soft decoration in this project are mainly in fresh and elegant color, shaping a contrast but harmonious relationship with reddish brown furniture. Thus a harmonious space was created to unify the styles of the whole space, community environment and interior decoration.

Most of the furnitures are American style furnitures with simplified lines, rough volume, natural material and elegant color and shapes. In accordance with the requirement for comfort, each of them reveals the natural smell of the sun, green grass and dew. Delicate and graceful fabrics with abstract plant pattern adorn the furniture, creating a free, warm and soft atmosphere, a typical American country style.

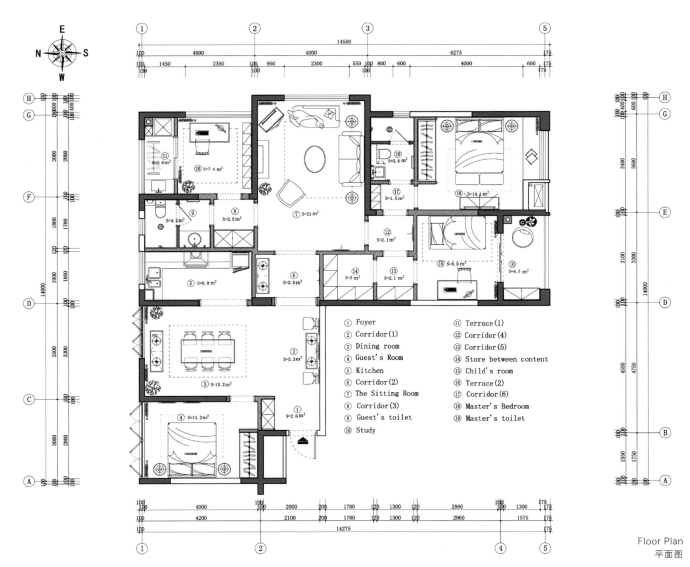

① Foyer
② Corridor (1)
③ Dining room
④ Guest's Room
⑤ Kitchen
⑥ Corridor (2)
⑦ The Sitting Room
⑧ Corridor (3)
⑨ Guest's toilet
⑩ Study
⑪ Terrace (1)
⑫ Corridor (4)
⑬ Corridor (5)
⑭ Store between content
⑮ Child's room
⑯ Terrace (2)
⑰ Corridor (6)
⑱ Master's Bedroom
⑲ Master's toilet

Floor Plan
平面图

作为居室来讲，色彩关系和谐是第一位的，不管采用什么样的配色方案，一定要取得和谐的基本效果。本方案的墙纸、天花板、软装多以清新淡雅的颜色为主，但配置红棕色的家具作为颜色的对比色，和谐的颜色构成了和谐的空间，使整个空间的风格、社区环境与室内装饰三者风格相统一。

本案的家具多为美式家具，有着简化的线条、粗犷的体积、自然的材质和较为清新淡雅的色彩及造型，但整体以舒适为设计准则，每一件都透着阳光、青草、露珠的自然味道，仿佛随手拈来，毫不矫情。有着抽象植物图案的清淡优雅的布艺点缀在美式风格的家具当中，营造出闲散与自在，温情与柔软的氛围，是典型的美式乡村风格。

European Style

欧式风格

Show Flat of Fuqing Jinhui Huafu

福清金辉华府样板房

Designer: Xiao Ming
Interior Design: North Coast Design
Location: Fuzhou, Fujian, China
Area: 160 m²
Photography: Zhou Yuedong

设 计 师：肖鸣
室内设计：北岸设计
项目地点：中国福建省福州市
面　　积：160 m²
摄　　影：周跃东

Keywords
关键词

European Style　欧式风格

Low-key Luxury　低调奢华

Active Space　空间活跃

Furnishings/Materials
软装 / 材料

Amasya Beige, Pan-American Wood Floor, European Furniture, Wallpaper

阿曼米黄、泛美木地板、欧式家具、墙纸

When entering the show flat of Fuqing Jinhui Huafu, the open space brings people a steadfast sense of tolerance, so people become easy to produce emotional resonance in the dialogue with the space. The designer gives this space a visual label of European amorous feelings; they do not intuitively file it up, but create it with inspiration.

In the functional areas of sitting room, dining room, study and bedroom, the wonderful coexistence of fashion and classic makes people feel grand and influential. The neo-classical style furniture and neatly strict layout fully show the unique European amorous feelings, and the building mass strewn at random creates rich space images from different angles. Placing oneself among them, people can spontaneously feel the heavy textured life and low-key luxury everywhere.

In addition to structure and furnishings, the details of the space cannot ignore the use of the light that makes the space living and bright. Overall, the main light source and point light source of this show flat jointly regulate the indoor light. They echo with each other by different attitudes to remind people of the stage and activate every corner of the room.

走进福清金辉华府样板房，开放式的空间呈现出令人踏实的包容感，在与空间的对话中很容易产生情感的共鸣。欧陆风情是设计师给予这个空间的视觉标注，但设计师并没有直观地去堆砌它，而是一种灵感的创造。

在客厅、餐厅、书房、卧室这些功能区域中，时尚与古典的奇妙共处，让人感受到一种名门气派的大家风范。新古典主义风格的家具以及工整严谨的布局，将风情独具的欧式情韵展现无遗，错落有致的体量感从不同角度营造出丰富的空间形象。置身其中，紧致厚重的生活质感油然而生，低调奢华的品位格调处处弥漫。

除了结构与陈设之外，空间的细节也不能忽视光线的运用，它是让空间鲜活起来的妙法。纵观全局，本套样板房里的主光源与点光源共同调节着室内的光线。它们以不同的姿态相互呼应着，让人联想起舞台的效果，一下子活跃了空间的每个角落。

Show Flat of Foshan Vanke Crystal City

佛山万科水晶城样板房

Designers: Shi Hongwei, Peng Zheng
Interior Design: Guangzhou C&C Design Co., Ltd.
Location: Foshan, Guangdong, China
Area: 110 m²

设 计 师：史鸿伟　彭征
室内设计：广州共生形态工程设计有限公司
项目地点：中国广东省佛山市
面　　积：110 m²

Keywords 关键词

European Style　欧式风格

Petty Bourgeoisie　小资情调

Bright Space　空间明亮

Furnishings/Materials 软装 / 材料

Crystal Product, Silver Product, Colored Glass Product, Carpet, Stone

水晶制品、银制品、有色玻璃制品、地毯、石材

This project applies white and silver as the main color tone and is adorned with blue and champagne gold, creating a life atmosphere full of petty bourgeoisie sentiment. Furniture with mellow and soft curved line, together with all kinds of simple decorations, reflects women's exquisite and delicate sense. It mainly uses crystal and silver accessories, and adds some colored glass, so these transparent materials make the space appear bright and rich with the layer feeling.

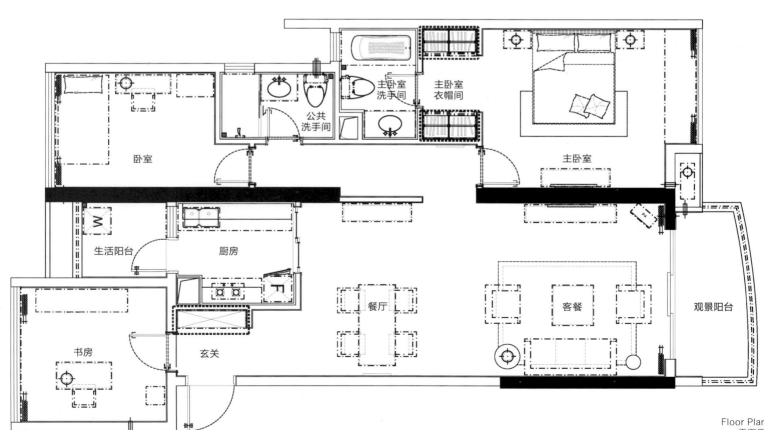

Floor Plan
平面图

本案整体以白色和银色为主色调，选用蓝色和香槟金色为点缀，营造一个富有知性小资情调的生活氛围，曲线柔美的家具和各种极具质感的饰品搭配，体现出女性般的细腻和精致感。饰品方面以水晶和银制品为主，搭配有色玻璃制品，通透的材质让空间显得明亮而富有层次。

Show Flat A of Greattown International, Fuzhou

福州名城国际 A 样板房

Designers: Lin Xinwen, Yu Yanqin
Interior Design: Hong Kong Pure Charm Design Co., Ltd.
Location: Fuzhou, Fujian, China
Area: 160 m²
Photography: Zhou Yuedong

设 计 师：林新闻　俞燕琴
室内设计：香港品川设计有限公司
项目地点：中国福建省福州市
面　　积：160 m²
摄　　影：周跃东

Keywords 关键词

European Style　欧式风格

Elegant and Gorgeous　儒雅富丽

Romantic　浪漫气息

Furnishings/Materials 软装/材料

Lightening Cream Marble, Magnolia Marble, Wooden Dado, Wallpaper, Mirror

闪电米黄大理石、白玉兰大理石、木面墙裙、墙纸、镜面

Gentle white color is utilized to present the extension and penetration of visual space, along with the spatial level variation, collocation of warm color elements and crystal lighting to create a space of Golden Rosa Rubus. Metal luster, glass transparence, stone thickness, leather texture and fabric warmth are collocated in various structures to present the extraordinary aesthetic sense and character under the golden lighting. With warm colors as main tone, the application of reflecting materials i.e. mirror, stone and metal strengthens the shading of light color, maintaining the consistence of spatial transparence and themed color. The golden atmosphere in general does not cheapen the space; instead, it brings the elegance and sumptuousness to the space with strong European-style colors space in splendid but not loudly way.

The space design is featured as romantic. Apart from the misted Western touch created by golden color, the shapes and radians of the furniture and lamps present deep French classical art. The background wall is mainly milky and brown with definite lines, leaving enough leeway for soft decoration. Sofa, tea table and

Floor Plan
平面图

dinettes are eye-catching with their unique European palace shapes and metal carving craft. Their symmetrical arrangement shows the spatial esthetics of classical European style. The moderate lighting reflects the European style's pursuit of detail and quality. The beautification of environment and decoration of interior space are enhanced as functions of lamps apart from their lighting function.

The design of bedrooms carries on the neoclassicism of the living room. Each single room has its own character on the basis the general style in pursuit of comfort as always. Large-area floor-to-ceiling glass windows, European-style palace beds, lamps and cabinets as the main structure, along with soft decoration of fabric, wallpapers and curtains in soft colors generate a room of warmth and tenderness. The gorgeous and romantic decoration brings the house splendid but not loudly quality, with sense of aesthetics in every detail.

在这套样板房中，设计师利用无侵略性的白色，展开视觉空间的延伸及穿透，并运用空间的层次变化、暖色系元素的搭配和水晶灯光的烘托，营造出"金色荼蘼"的丰富空间。设计师将金属的光泽、玻璃水晶的通透、石材的厚实、真皮的质感和布艺的温暖通过各种结构方式搭配出来，在金色灯光的映衬下，空间显示出不同寻常的美感与质地。空间以暖色系为主旋律，镜面、石材、金属等反光材料的运用强化了灯光色调的晕染，保证空间的通透感和主题色彩的一致。整体的金色氛围并没有使空间显得俗气，反而在带有浓烈欧式色彩的儒雅富丽中，将空间表现得华而不俗。

空间的设计颇具浪漫气息。抛开金色所营造的迷离般的西洋情调不说，单是家具、灯具的造型与弧度就都弥漫着浓浓的法式古典的艺术气息。背景墙的色调以奶白色和棕黄色为主，墙体线条明朗，为软装提供了充分的发挥余地。沙发、茶几、餐桌椅以其独有的欧式宫廷造型和金属雕花工艺来吸引观者的眼球，对称的摆设也流露出古典欧式风格的空间美。欧式风格对于金属和水晶灯的要求从未降低，恰到好处的灯光辉映来源于细节打造和对质感的追求。灯具在本案的传统照明功能并没有弱化，对美化环境和装饰室内空间的功能却在加强。

卧室则延续了客厅新古典主义的情调，每个单间在保持风格的基础上各具风情，崇尚的依然是一如既往的舒适。大面积的落地玻璃窗、欧式宫廷床、灯饰和壁柜为结构主体，布艺、壁纸、窗帘等软装皆选用暖色系，营造出一室的暖意和温情。华美浪漫的装饰让家居有着华而不俗的气场，美感的流露存在于每一个细微之处。

Show Flat B of Greattown International, Fuzhou

福州名城国际 B 样板房

Designers: Lin Xinwen, Lin Jin
Interior Design: Hong Kong Pure Charm Design Co., Ltd.
Location: Fuzhou, Fujian, China
Area: 180 m²
Photography: Zhou Yuedong

设 计 师：林新闻　林锦
室内设计：香港品川设计有限公司
项目地点：中国福建省福州市
面　　积：180 m²
摄　　影：周跃东

Keywords 关键词	European Style　欧式风格 Simple Element　简约元素 Elegance　高贵大气
Furnishings/ Materials 软装/材料	Alligator Skin, Leather, Flannelette, Satin, Wood Veneer, Archaized Marble, Navona Travertine 鳄鱼皮、皮革、绒布、缎布、木饰面板、仿古大理石、罗马洞石

Generally speaking, the furnishing of classical style is able to present the quality of elegance, splendor and grace in a house with enough area, which is the reason why European style is popular. In the show flat, designers underline the spatial elegance and splendor through European neo-classical style, with integration of diversified simple elements manifesting the low-key luxury. Sedate colors, transparent space, exquisite details and linear shapes bring the show flat gentleness and romance out of splendor.

The space of living room is the one requires most time and energy from designers, as well as the key part of the general style. A large area of floor-to-ceiling glass windows is set in living and dining rooms without any complex partition. The spatial structure attains the classical aesthetics of solid wood materials, and modern materials i.e. glass and metal etc. Wooden frames are used to cut the walls into several pieces of geometric plates, building the classical Western style. Carrying on the comfort sense of neoclassicism, the general space is mainly in warm colors collating with fabric sofa and tea tables of carved metal in cold colors, creating sense of serenity. The dwelling lifestyle is much more elegant and graceful for the sense of mixed

Floor Plan 平面布置图

texture and abundant sense of levels. The designers highlight the level sense of the space through the design of walls; either carved texture or hanging painting and calligraphy on the walls could underline the entire space through their luxurious decoration and strong colors.

The designers have arranged four feature romantic bedrooms. Sharing the same filling elements and major tones, the bedrooms are distinguished from each other through the variation of colors, materials, shapes and layout, under the color theme of red, brown, white and silver respectively. Bedclothes in warm colors with various kinds of lamps, side tables, curtains, bolsters generate sense of warmth and gentleness, as well as the distinguished classism and luxury. Particularly every room is set with bay window or French window for views on vast river outdoor, simple but sumptuous.

通常来说，足够的住房面积可以让古典风格的家居尽情展示其大气、华丽、高贵的气质，这也是欧式风格之所以受大户青睐的重要原因。在这套样板房中，设计师以欧式新古典主义风格凸显空间的堂皇与奢华之美，同时也融合了多元化的简约元素，展现居室低调的华丽。沉稳的色调、通透的空间、精致的细节与线条造型，让这套样板房不只是富丽堂皇，也拥有温婉、多情的一面。

客厅装修无疑是设计师花费时间和精力最多的空间，也是整个居家格调的重墨之笔。设计师以大面积的落地玻璃窗为起始，客厅、餐厅之间没有复杂的隔断，空间框架继承了实木材料的古典美，玻璃、金属等现代材质也被运用其中，一个个木质框架将四周墙壁分割成若干几何板块，塑造了经典的西式风格。设计师还秉承了新古典主义一如既往的舒适，空间整体以暖色调为主，搭配冷色系镶雕花金属的布艺沙发和茶几，增添了几分沉静，充满质感的混搭和丰富的层次中所包容的大度使居室品质更具大家风范。设计师通过对墙体的刻画凸显空间的层次感，背景墙无论是雕花纹理还是书香挂画，都能通过华丽的装饰与浓烈的色彩烘托出整个空间的氛围。

设计师精心布置了四间极具浪漫特色风情的卧室。在填充元素及主体基调不变的基础上，通过色彩、材质、造型、布局改变，以及红、棕、白、银四个主题色调的风格微调，展现了百变的华丽。床品皆选用暖色系，搭配各种灯饰、边几、窗帘、抱枕，营造出暖意和温情，显得格外的古典和奢华。值得一提的是，每一个房间都设置了飘窗或落地窗，抬眸一望，室外的浩瀚江景一览无余，将窗外景色引入室内，简单却又不失华贵气息。

Type D Show Flat of Qinyang Rose City

沁阳玫瑰城 D 户型样板房

Designers: Zhen Chengyu, Gao Shan, Cen Yun
Interior Design: Hanzhou 5 Pen Architectural Decoration Design Co., Ltd.
Location: Qinyang, Henan, China
Area: 160 m²

设 计 师：郑成余　高杉　岑云
室内设计：杭州五朋建筑装饰设计有限公司
项目地点：中国河南省沁阳市
面　　积：160 m²

Keywords 关键词

Simplified European Style　简欧风格

Cozy and Lofty　舒适典雅

Space Continuity　空间连续

Furnishings/Materials 软装／材料

Carpet, Veiling, European Style Wallpaper, European Style Furniture, Classical Decorative Painting

地毯、帐幔、欧式墙纸、欧式家具、古典装饰画

In terms of plane design, the designers change a bedroom into a dining room and a study, which improve the functions of the room and keep the integrity of the overall layout of this hose. The hallway is in the corridor of the bedroom that seems to be not too long visually so as to avoid the toilet faceing the entrance directly. As for the main materials, it is mainly dark color wood furnishes, European style wallpaper and European line, complemented with cozy and lofty soft decoration and simplified European style furniture to build up a safe and sound simplified European space.

Inheriting traditional European style, the simplified European style absorbs its characteristics. The design chases for the continuity of spatial change and the layering of the transforming between the feeling and forms. Mostly the pattern wallpaper, carpet, curtain, bedspread and classical decorative painting display a luxury style. The decoration around the line part of frames is used with lines or golden lines, which stand

out the concave-convex feeling and give elegant arc lines in terms of appearance design. The design philosophy is to pursue such a representation that nobility presents in deep and the lofty is soaked with luxury. That is expected to fully display the owner's life attitude that considers the life as an art and look for a high-quality and lofty life.

　　在平面设计方面，设计师将一间卧室改为餐厅和书房，这样既完善了房间功能，又不破坏房间整体格局的完整性。在卧室的过道处设置玄关，使得视觉上不会感觉过道过长，也避免了卫生间门正对入户门的尴尬。主材方面以浅色木饰面、欧式墙纸及欧式线脚为主，配以舒适典雅的软装和简欧家具营造出稳重的简欧空间。

　　简欧风格继承了传统欧式风格的装饰特点，吸取了其风格的"形神"特点。在设计上追求空间变化的连续性和形体变化的层次感，室内多采用带有图案的壁纸、地毯、窗帘、床罩、帐幔及古典装饰画，体现华丽风格。画框的线条部位装饰为线条或金边，在造型设计上既要突出凹凸感，又要有优美的弧线。其设计哲学都是追求在深沉里显露尊贵、典雅中浸透豪华的设计表现，并期望这种表现能够完整地体现出业主追求品质、典雅生活，并视生活为艺术的人生态度。

Show Flat of Shenzhen CTS International Mansion

深圳中旅国际公馆样板房

Designer: Gong Decheng
Interior Design: Shenzhen Gong Decheng Interior Design Office
Location: Shenzhen, Guangdong, China
Area: 160 m²

设 计 师：龚德成
室内设计：深圳市龚德成室内设计事务所
项目地点：中国广东省深圳市
面　　积：160 m²

Keywords 关键词

European Style　欧式风格
Elegance and Luxury　典雅奢华
Nobility　高贵气质

Furnishings/Materials 软装/材料

Apollo Marble, Arabesecato Faniello, Imported Wallpaper, Handmade Carpet, Woodcarving, Tawny Glass With Grinding Flower

阿波罗大理石、特级大花白、进口壁纸、手工地毯、木雕花、磨花茶镜

The elegant space brings peace to one's heart, while the artistic environment influences the characteristic and temperament of the householders. The classical atmosphere is the essence of the project.

The sitting room is decorative with modern European style. The pale pinkish purple suede sofa presents the elegance and gorgeousness of European style. The particular materials for the overall design imply the sense of luxury without poor taste. The European tea table with silver edge is dotted with flower carpet, which adds lively atmosphere for the entire space. The light of the crystal chandelier brings nobility to the entire sitting room. The combination of flowers and black gauze on the curtain is elegant and dynamic at the same time, presenting hazy feeling for the house.

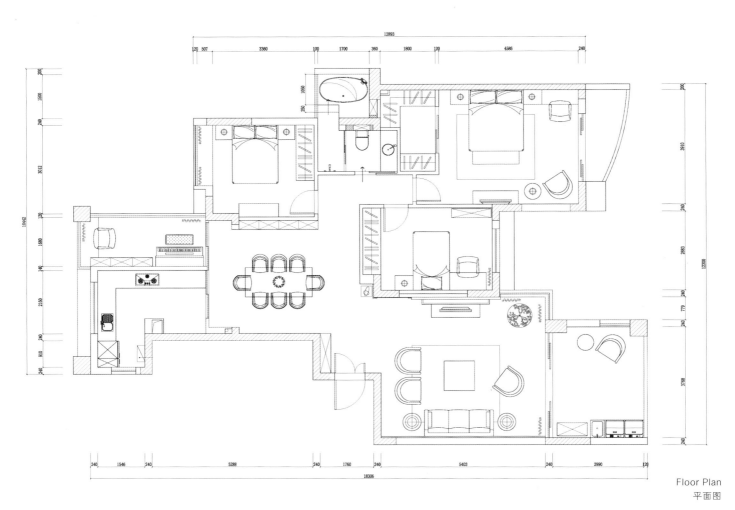

Floor Plan
平面图

On the design for the canteen, the comparisons between deep and light, cold and warm are used to express the modest luxury, and display the noble quality out of simplicity in the canteen.

The bathroom is simply designed with marble as the main material, which is convenient to clean. The arced groove on the wall is the highlight of the bathroom design, which is used as a small storage shelf.

The purple suede bed in the master bedroom is elegant and fantastic. The furniture are mainly in black with silver edge, elegant in implicit. The frosted glass with decorative patterns replaces the traditional door, giving more sense of transparency and decoration to the

house. The soft roll wall softens the bedside, while the color of brown brings the best of the implicit design with high quality.

The flower-pattern wallpaper in the secondory bedroom makes the steady room much lively. The white wardrobe and the silver bedside table form the obvious comparison. One is simple and implicit and the other is noble and steady, avoiding dull feeling in the bedroom.

Children's simplicity is attached to the children's bedroom with colorful wallpaper and bedclothes with cartoon patterns. The black-and-white curtain and the white cabinet follow the designer's tradition again: low-key steadiness and noble elegance.

优雅迷人的空间，可以沉静涤清繁杂的心灵，而充满艺术的绝佳环境，更能潜移默化影响业主的个性与气质。洋溢古典气质的氛围，这就是本案设计的精髓之处。

客厅以后现代欧式风演绎，藕荷色的绒面沙发完美展现了欧式风的雍容华丽，整体设计用材考究，低调中隐隐透着奢华感，又不具俗气。银边镶嵌的欧式茶几下面以花朵地毯作为点缀，红色的花朵在黑色背景下娇艳盛开着，为整个空间增添了鲜活的气息。水晶吊灯散发着朦胧的光芒，且透出的高贵气息渲染了整个客厅。窗帘上花朵与黑纱的结合散发出高贵气息的同时也不失生动感，为居室平添了些许朦胧感。

在餐厅的设计上，以深与浅、冷与暖的对比来诠释低调的奢华感，一面突出一面衬托，这样的设计使得整个餐厅简洁中尽显高贵气质。

洗手间的设计以简洁质朴为主，大理石为主材料，方便清理同时兼具装饰性。墙壁上的圆拱凹槽是整个洗手间的亮点，设计师把它设计成小型的储物架，个性十足。

主卧紫色的绒面床散发着高贵梦幻的气息，家具以黑色为主色系，银边镶嵌，低调内敛中透着些许华丽典雅。以花纹式磨砂玻璃的推拉门代替了传统的门，使得居室更具通透感和装饰感。床头软包墙面以软包材料体现柔性美，咖啡色把沉稳内敛发挥到极致，大气又不失质感。

次卧花朵图案的壁纸使得这个沉稳的卧室多了几分俏皮。白色的衣柜与银色的床头柜形成鲜明对比，既简洁内敛，又高贵沉稳，这样的动静组合让这个简单的卧室不觉单调。

设计师以彩色条纹壁纸和卡通图案的床具为儿童房注入了童真感。黑白图案的窗帘和白色的柜子又秉承了设计师一贯的传统——低调沉稳中尽显高贵典雅。

Show Flat of Vanke Wonder Town, Chongqing

重庆万科缇香郡样板房

Designer: Danfu Liu
Interior Design: PINKI Creative Group and Liu & Associates (IARI) Interior Design Co., Ltd.
Furnishing Design: PINKI ZBJ Design Institute
Location: North New Area, Chongqing, China
Area: 300 m²

设 计 师：刘卫军
室内设计：PINKI（品伊）创意机构＆美国IARI刘卫军设计师事务所
陈设设计：PINKI知本家陈设设计机构
项目地点：中国重庆市北部新区
面　　积：300 m²

Keywords 关键词

Neo-classical European Style　欧式新古典风格
Noble and Elegant　典雅高贵
Dignity　舒展大方

Furnishings/Materials 软装/材料

Jazz White Marble, Black Woodgrain Stone, Blue Emulsion Paint, Wallpaper, Tempered Glass, Elm White Veneer
爵士白大理石、黑木檀纹石、蓝色乳胶漆、墙纸、钢化玻璃、榆木索白饰面

The design is concise but not simple by retaining the classics and abandoning the drosses. All the details are carefully designed to achieve an elegant and noble effect.

In the living room, elegant light blue, light purple and jazz white show the female romance, while nostalgic dark brown, dark red and black remind people of the gentlemen. The Chinese-style kitchen with diamond-shaped tiles, the dinning room with good wines, and the elegant western-style kitchen enable the client to have more dining choices. In the master bedroom, lace beddings and curtains, crystal chandelier of black rose color, dark-red nightstands and silver-red bed stool, all show a unique style and unspoken romance.

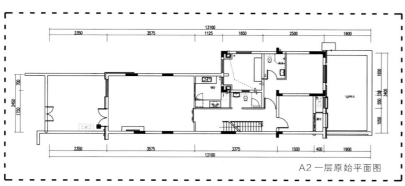

A2 一层原始平面图

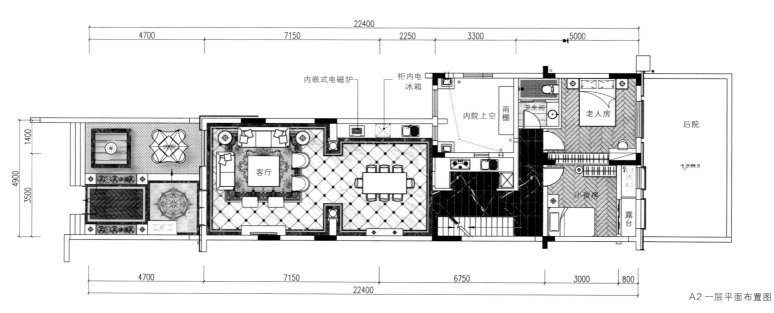

A2 一层平面布置图

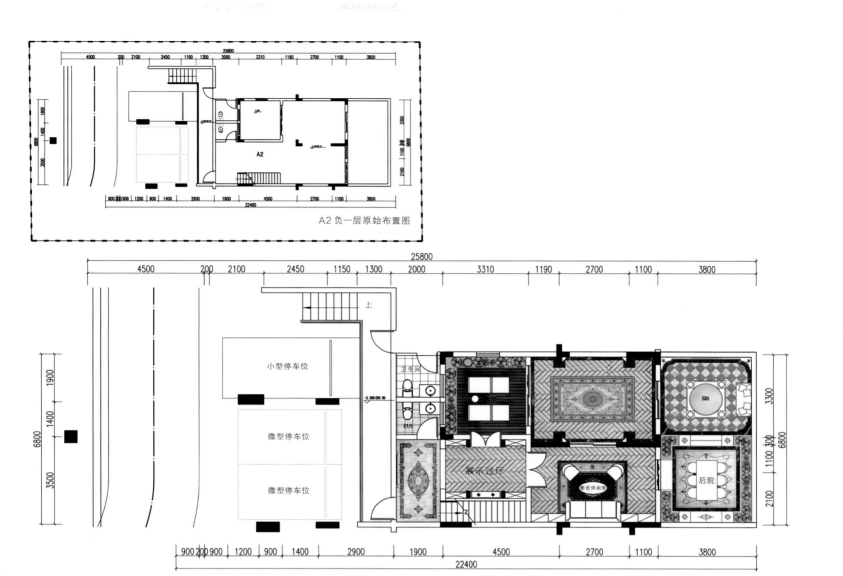

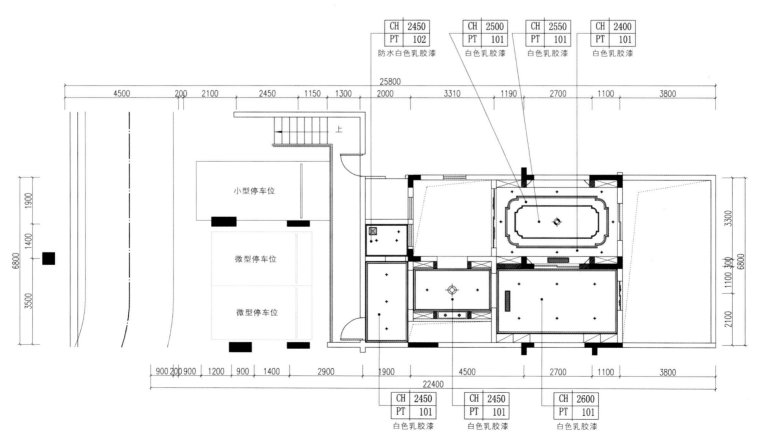

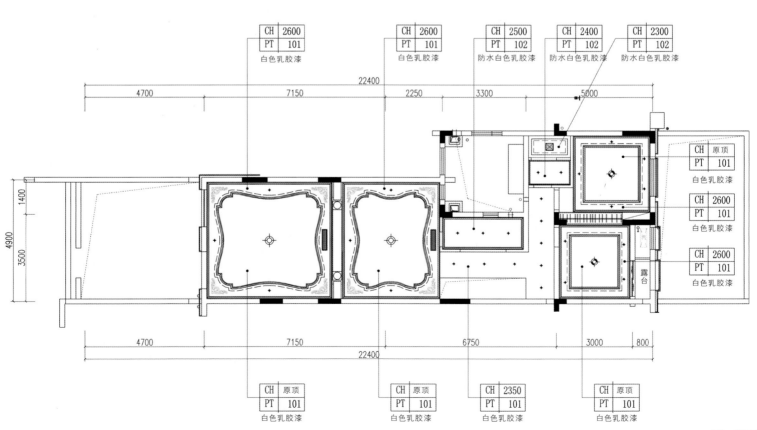

A2 一层天花布置图

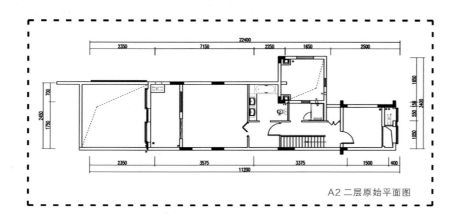

A2 二层原始平面图

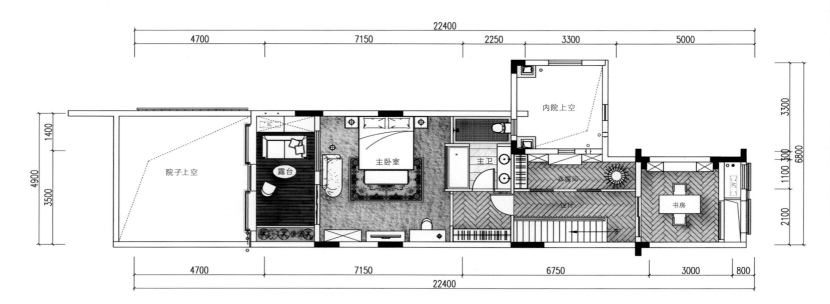

A2 二层平面布置图

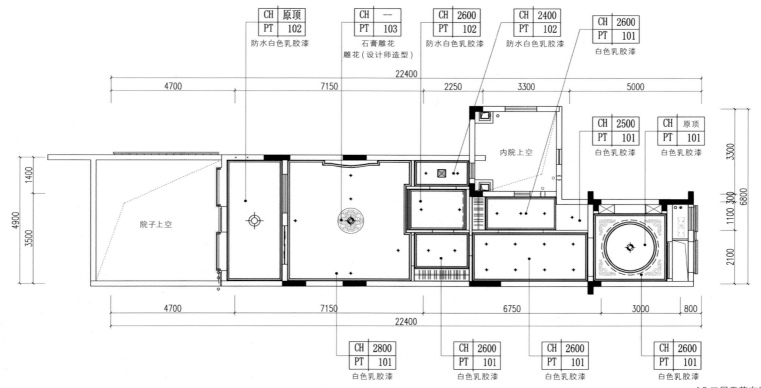

A2 二层天花布置图

本案的整体设计简约而不简单，去掉腐朽，保留经典，从整体到局部，精雕细琢，镶花刻金，典雅而高贵。

客厅里，优雅的淡蓝、浅紫、爵士白，如同女性的浪漫情愫；怀旧的深啡、暗红和黑，如同男性的绅士情怀。菱形花砖的中式厨房与餐厅里摆放着美酒、舒展大方的西式厨房让主人可以尽享中西餐饮。主卧床幔帘头花边、黑玫瑰色的水晶宫灯、暗红床头柜、银绒红床尾凳的质感展现出独特的气质，言说浪漫，亦不过如此。

Type B Show Flat of Chongqing Zhonghai International Community No.3 Building

重庆中海国际社区 3 号楼 B 户型样板房

Designer: Danfu Liu
Participating Designers: Zhong Wenping, Liang Yi, Zhang Luogui, Liu Shumiao, Chen Liqiang
Interior Design: PINKI Interior Design and Liu & Associates (IARI) Interior Design Co., Ltd.
Furnishing Design: PINKI Art & Deco Design
Developer: China Overseas Holdings Limited (Chongqing)
Location: Nan'an District, Chongqing, China
Area: approximately 138 m²

设 计 师：刘卫军
参与设计：钟文萍　梁义　张罗贵　刘淑苗　陈利强
室内设计：PINKI（品伊）创意机构＆美国IARI刘卫军设计师事务所
陈设设计：PINKI（品伊）创意机构＆知本家陈设艺术机构
开 发 商：中海地产重庆有限公司
项目地点：中国重庆市南岸区
面　　积：约138 m²

Keywords 关键词

European Style　欧式风格
Modern Fashion　时尚现代
Dynamic Line　动感流线

Furnishings/Materials 软装／材料

White Oak Veneer, Wallpaper, Jazz White Marble, Oak Floor, etc.

白橡木饰面、墙纸、爵士白大理石、橡木地板等

"Fashion, elegance and inspiration" is the theme of the design. Due to the small area, the designers choose light colors for this house to create a magic haven of modern style in this hustling and hustling city. It keeps the original layout and pays attention to the connections between different spaces.

By using different ceiling materials, the designers have created different space atmosphere in the dining room and living room. The mirror on the ceiling of the dining room makes the space higher visually. Modern fashionable design combines with the European style chandelier to enhance the sense of quality.

The design is exaggerated and bold to use large-area jazz white marbles. In addition, with some petit bourgeoisie sentiments, it makes the space feel warm and comfortable. The passage will leads people into the master bedroom, giving the experience of intimacy and romance.

Plan of Type B
B 户型面布置图

  "时尚、优雅、灵感"是本案设计的主题。本案户型不算大，整体色调以浅色调为主。设计师以简洁现代的设计思想为指导，致力将这里打造为繁华都市里的神奇"避风港"。设计师对整个户型保留完整的空间划分，注重对室内空间的流线优化。

  在餐厅与客厅为同一空间的情况下，设计师运用天花板材质的区分，营造不同空间的感受。餐厅天花板的镜面设计，将空间在感官视觉上拉高。时尚现代的造型，搭配欧式古典元素的水晶吊灯使空间更具气质。

  设计师以夸张大胆的设计手法，将公共空间大面积使用爵士白大理石的同时也将现代都市新贵的小资情结融入其中，使空间更为温馨舒适。步入通向主人私密空间的通道，走进卧室，倍感温情与浪漫。

Fuzhou East Great Town Show Flat

福州东方名城样板房

Designer: Li Xiwei
Approval: Ye Bin
Designed by: Fujian Guo Guang Yi Ye Decoration Group
Area: 200 m²

设 计 师：李西威
方案审定：叶斌
设计单位：福建国广一叶建筑装饰设计工程有限公司
面　　积：200 m²

Keywords 关键词

Classical European Style　欧式古典风格

Dignified & Graceful　雍容华贵

Concise & Lively　简练明快

Furnishings/Materials 软装/材料

Marble, Glass, Gold Foil, Stainless Steel

大理石、玻璃、金箔、不锈钢

This case is a flat in European style with florid ornament, strong colors and exquisite shape. In addition, designer use gorgeous droplights to create atmosphere and large glass window to bring in favorable daylighting, which makes it much more stylish. Velvet and fabric sofa units have beautiful well-crafted curve that combines the luxury of traditional European houses and utility function of modern houses perfectly. Doors for the rooms and cabinets are designed in two different styles with elegant arc lines, appear fascinating and charming. Besides the lively, romantic and abstract European style, one may also feel the pursuit of nature, a tranquil, free lifestyle in this flat.

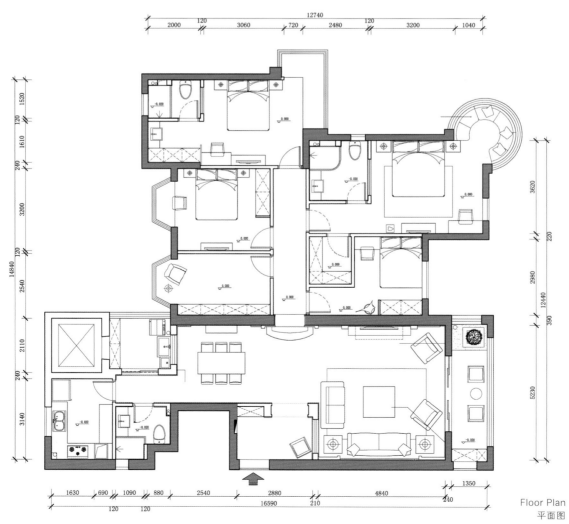

Floor Plan
平面图

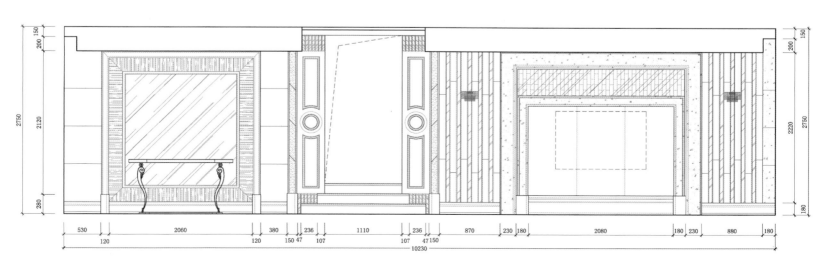

Elevation 1
立面图 1

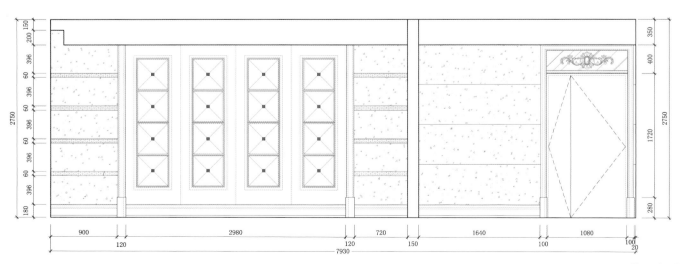

Elevation 2
立面图 2

本案属于欧式风格，设计师以华丽的装饰，浓烈的色彩，精美的造型营造出一种雍容华贵的装饰效果。设计师采用华丽的吊灯营造气氛，大面积的玻璃窗带来了良好的采光，落地的窗帘显得更加气派。丝绒的布艺沙发组合，有着美妙的质感曲线，将传统欧式家居的奢华与现代家居的实用完美地结合了起来。门的造型设计，包括房间的门和各种柜门，既突出凹凸感，又有优美的弧线，两种造型各有特色，风情万种。整个空间在追求简练明快、浪漫、单纯、抽象的欧式风格时，还流露出对自然的向往，表达了一种恬淡、洒脱的生活方式。

Show Flat of Sansheng Central Park

三盛中央公园样板房

Designer: Ye Meng
Program Validation: Ye Bin
Interior Design: Guoguang Yiye Decoration Group
Location: Fuzhou, Fujian, China
Area: 400 m²
Photography: Ye Meng

设 计 师：叶猛
方案审定：叶斌
室内设计：福建国广一叶建筑装饰设计工程有限公司
项目地点：中国福建省福州市
面　　积：400 m²
摄　　影：叶猛

Keywords 关键词

European Classicism　欧式古典风格

Dignity & Sumptuousness　尊贵奢华

Fashion　时尚气息

Furnishings/Materials 软装/材料

Archaized Tiles, Stone, Mirror, Gold Foil

仿古砖、石材、镜面、金箔

The classicism after baptism is re-defined in this project and the strong sense of design and sumptuousness are exacted and integrated into the modern architecture. To manifest the sense of sumptuousness and fashion, the design language of the style is applied to the space with the designer's master of European elements and new luxury style. The space of fashion atmosphere and European Classicism has expressed the elegant and graceful quality in an unobtrusive way.

经过洗礼的古典风格在这里被再次定义,极强的设计感和奢华感被再次提炼,融入到现代建筑中。为了体现奢华的时尚感,设计师通过对欧式元素的把握和对新奢华风格的理解,将体现这些风格的设计语言运用到空间中来,使空间充满时尚气息,使优雅从容的气质毫不张扬地在这个欧式古典风格的空间中一一呈现。

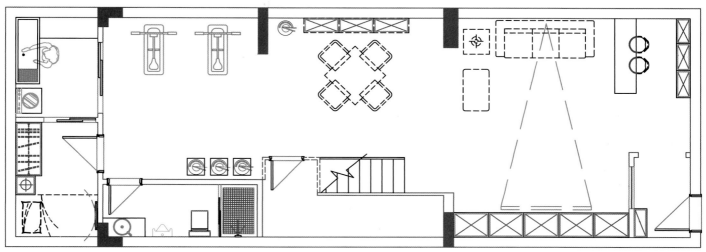

Basement Plan
地下层平面图

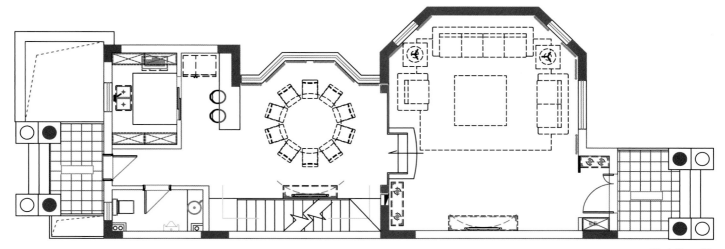

First Floor Plan
一层平面图

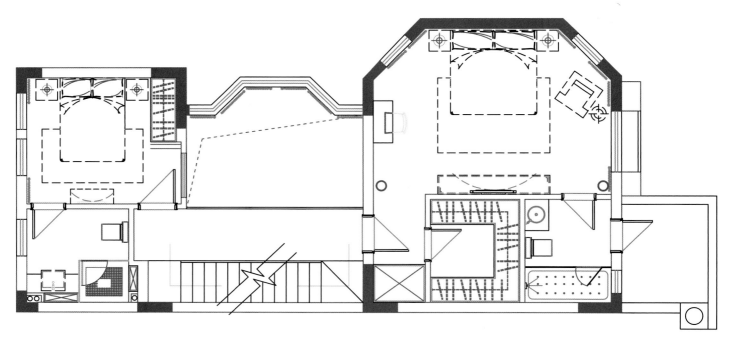

Second Floor Plan
二层平面图

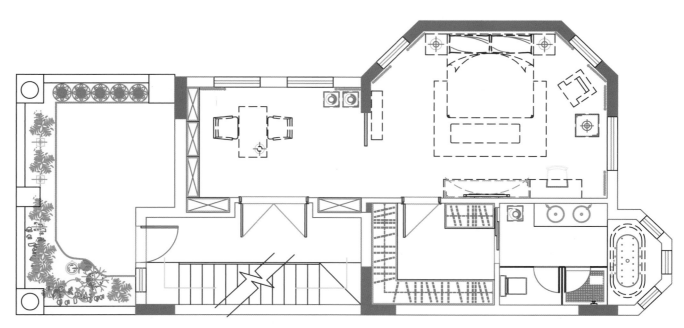

Third Floor Plan
三层平面图

Show Flat of Forte Ronchamps, Nanjing

南京复地朗香样板房

Designer: Fabrizio de Leva
Interior Design: FDL Architects
Location: Nanjing, Jiangsu, China

设 计 师：法布里奥·德·莱瓦
室内设计：FdL建筑与设计事务所
项目地点：中国江苏省南京市

Keywords 关键词

Italian Style　意大利风格

Exquisite & Luxury　精致奢华

Spaciousness　空间宽敞

Furnishings/Materials 软装/材料

Ceiling Lamp, Wallpaper, Floor, Stone

吊灯、墙纸、地板、石材

The project is designed under the following concept: exquisite details of traditional sumptuousness, sense of history and art in space decoration, furniture and furnishings of modern and fashion elements manifesting modern lifestyle in space embellishment; tradition and fashion as two design languages are adopted to build the delicate and sumptuous life space of moderns. The integration of two elements presents not only the dignified temperance of the project, but also the authentic modern lifestyle of Italy to the owner.

The designer adopts the integration of classical hard decoration and modern furnishing based on the understanding of domestic clients and actual market, so as to lead the pure Italian lifestyle into the product. The design also creates a strong visual conflict and a comfortable and luxurious life space through the collocation of applicable colors and materials. The dignified classical hard decoration and the comfortable modern furnishing reflect a new dwelling lifestyle.

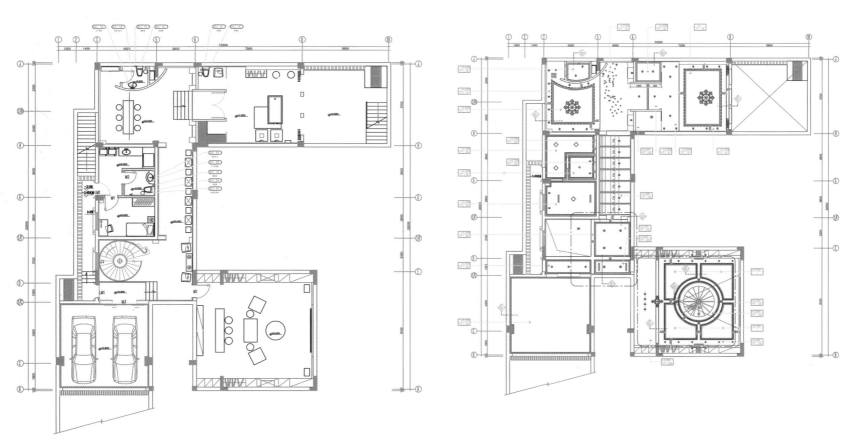

Basement Floor Plan
负一层平面布置图

Ceiling Plan of the Basement Floor
负一层天花布置图

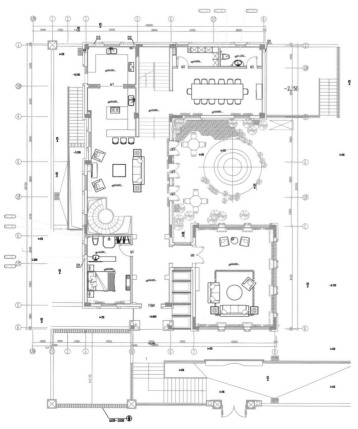

1st Floor Plan
一层平面布置图

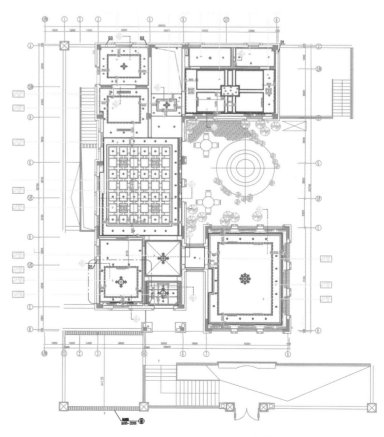

Ceiling Plan of the 1st Floor
一层天花布置图

The designer takes full consideration about the rational function layout. The coherence and convenience of public space provides a reasonable and comfortable streamline way: from entrance hall to the living room, and to the kitchen and dining room, going downstairs from the end stairs to the private club area; the entire streamline is smooth. Speaking of private space, the aged room in the first floor considers the move difficulty of the aged and security; the children room in the second floor provides relatively independent space; the main bedroom in the third floor occupies a more private spatial layout and shows the dignity of the owner. The study in the fourth floor as a way of temperance improvement, offers the man of the house a quiet, comfortable and graceful space for reading and work.

  本案的设计理念就在于：将传统奢华、富有历史感和艺术感的精致细节用到装饰空间上，将富有时尚现代元素、体现现代生活方式的家具软装来点缀空间，用传统与时尚两大设计语言打造属于现代人的精致奢华生活空间。这两种元素的融合，成就了本案高贵不凡的气质，也在业主面前呈现了原汁原味的意大利当代生活方式。

  设计师根据对国内客户的了解，从实际的市场出发，大胆采取了古典硬装和现代软装结合的设计理念，将纯正的意大利生活方式引入该产品中，并且使用强烈的视觉冲突，通过合适的色彩搭配以及材料搭配，营造出一种舒适奢华的生活空间。硬装古典厚重的基调，加上现代舒适的家具和软装，无不展现一种全新的家居生活方式。

  设计师也充分考虑了功能布局的合理性。公共空间的连贯性和便利性给住户提供一种合理舒适的流线方式，从入口玄关到客厅，再到起居室、厨房、餐厅、从尽头的楼梯下楼便可以抵达私家会所区域，整个流线非常顺畅；从私人空间上来讲，一层的老人房考虑了老人行动不便的特点以及行动的安全性；二层小孩房为自己的孩子提供了相对独立的空间；三层的主卧空间能够让空间布局更为私密，同时也体现了主人的尊贵；而四层的书房提升了本案的品质，同时给男主人提供了一个安静、舒适、典雅的阅读、办公空间。

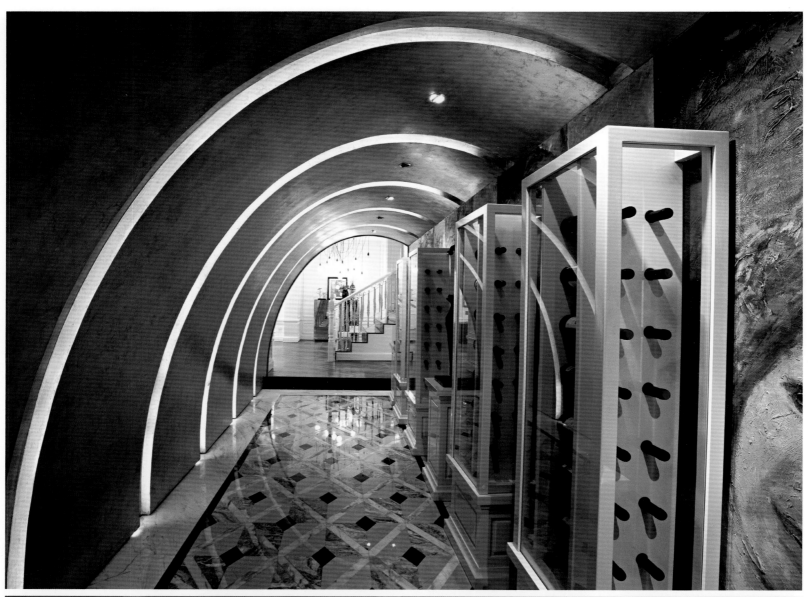

Hung Hom Lingshijun Pulin Garden — Showflat of Apartment A

红磡领世郡普霖花园 A 户型样板房

Designers: Chen Yi Zhang Muchen
Interior Design: Fenghe Muchen Space Design
Location: JinNan, Tianjin, China
Area: 620 m²
Photography: Zhou Zhiyi

设 计 师：陈贻 张睦晨
室内设计：风合睦晨空间设计
项目地点：中国天津市津南区
面　　积：620 m²
摄　　影：周之毅

Keywords 关键词	European Style　欧式风格 Castle Flavor　古堡气息 Elegant and Profound　优雅深沉
Furnishings/ Materials 软装/材料	Solid Wood Flooring, Wallpaper, Mosaic, Emulsion Paint 实木地板、壁纸、马赛克、乳胶漆

After the designers' repeated scrutiny, the whole space applies elegance and profoundness as the overall tone for the space, using pure English leather furnitures and pure copper droplight to foil out the overall atmosphere of the space, and the skillful application of the design techniques highly explains the rational life atmosphere of noble temperament. The color mainly adopts stable & deep brown series, beige series and warm color tones. The decorative surface uses modern modeling method combined with Britain traditional language that it chooses brown leather materials, warm color of stone materials and solid wood material to try to create low-key luxury and high-quality living space.

The sitting room is a 7 m high space, and the designers boldly introduce Britain pure tall gothic rose window into the space to form an adornment focus. The application of rose window brings wonderful decorative effect to the space, adds the space with more symbolic meanings, and creates strong European culture

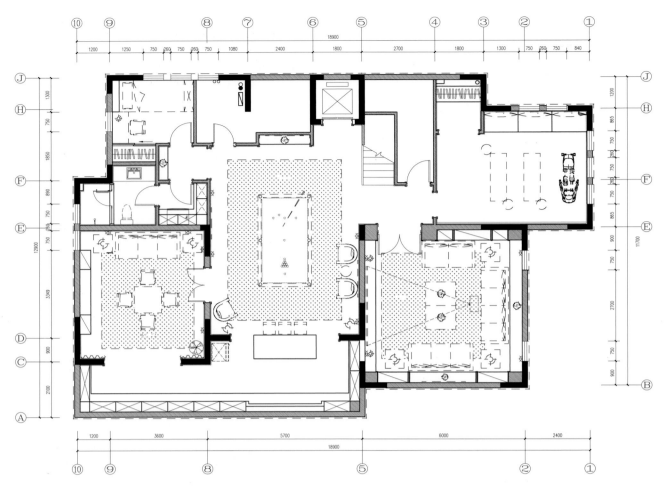

Plan for Basement Floor
地下一层平面图

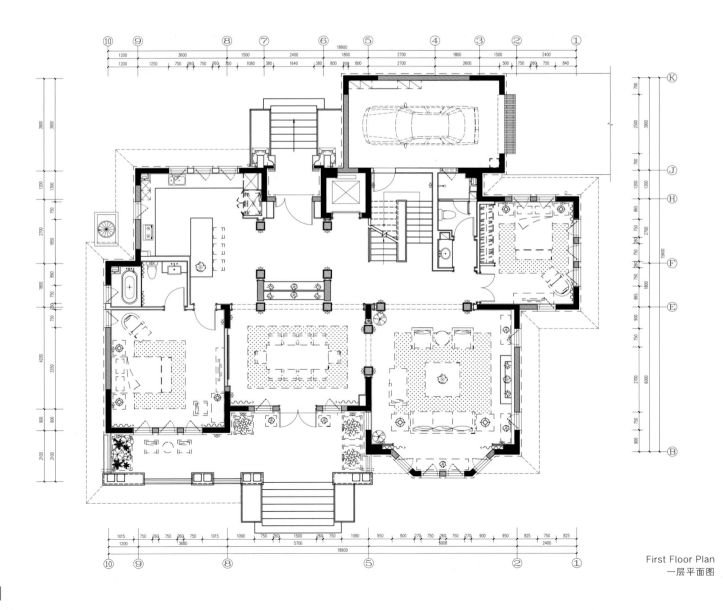

First Floor Plan
一层平面图

breath for the overall space, which makes it beyond the concept of "popular", but become the high-end symbols of status and taste.

The villa has a total of four floors, and the design style of each floor echoes with the overall temperament, thus brings people feeling similar to the Britain castle mysterious breath that makes the whole space solemn and grave but does not break the romantic sentiment. This element of rose is also used on the background wall of the restaurant and the underground recreation room; the circular contour of the rose window and internal structure give out a centripetal appeal, which forms a visual focus, highlights the visual recognition features of the space, and also fills the whole design with unique spiritual temperament and spiritual focus.

经过设计师反复的推敲，整个空间以优雅深沉为空间整体基调，使用纯牛皮英式家具以及纯铜吊灯烘托空间整体气氛，设计手法的纯熟运用充分诠释着贵族气质的理性生活氛围。整体色调以深沉稳重的褐石色系和米色系等暖色调为主，装饰面通过运用现代造型方式结合英式的传统形式语言，选用棕色系真皮材质、暖色系石材以及实木饰面等材质，在空间中尽力营造出一种低调奢华的高品质居室氛围。

客厅为挑空7 m的高空间，设计师将英国纯正高挑的哥特式玫瑰花窗大胆地引入此空间，形成装饰重点。玫瑰花窗的运用给空间带来奇妙的装饰效果，同时也带给了空间更多的象征意义，并且为整体空间营造出了浓厚的来自欧洲的文化气息，使其超越"流行"概念，而成为一种地位和品位的高端象征。

该别墅一共有四个楼层，每个楼层风格处理均与整体气质相呼应，带给空间感受者类似英式古堡略显神秘气息的立体感受，让整个空间肃穆庄重而不失浪漫情调。玫瑰花窗这一元素的运用同样出现了餐厅以及地下空间娱乐室的背景墙上，玫瑰花窗圆形的外轮廓与内部的结构赋予了一种向心的吸引力，形成了视觉上的焦点，更是突出了空间的视觉识别特征，也让整个设计充满独特的精神气质和精神指向。

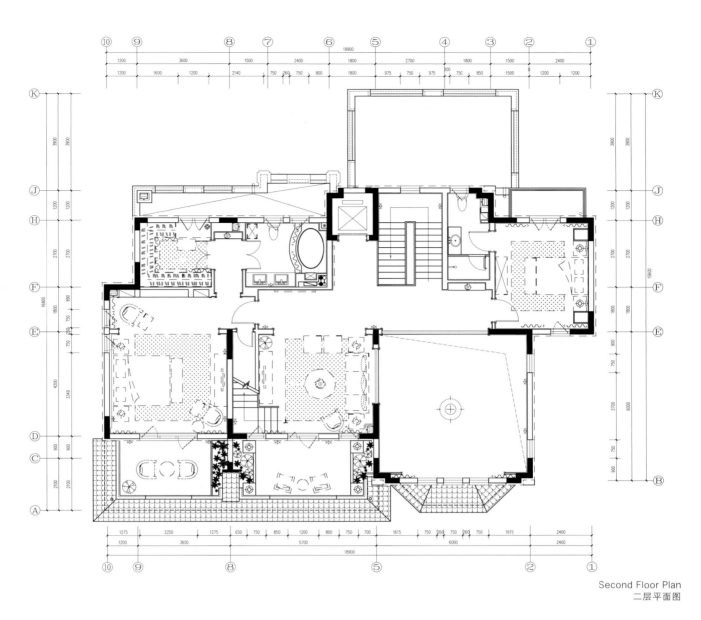

Second Floor Plan
二层平面图

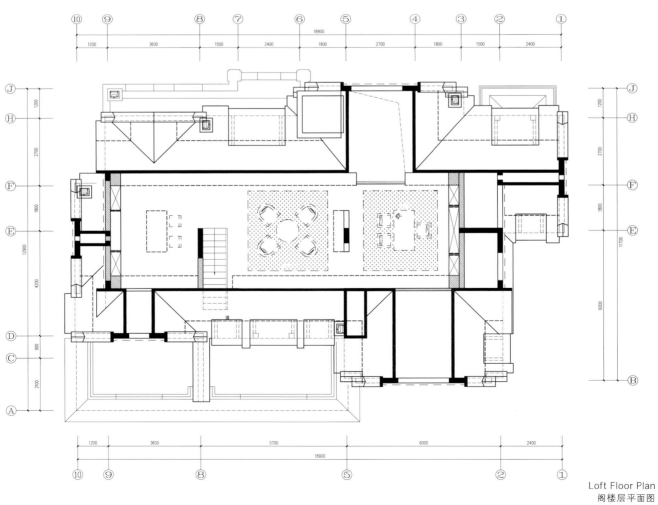

Loft Floor Plan
阁楼层平面图